I Am Not
Myself
These Days

A Memoir

JOSH KILMER-PURCELL

HARPER PERENNIAL

NEW YORK • LONDON • TORONTO • SYDNEY

This is a memoir of a certain time in my life. All of the names of major characters and other identifying details have been changed and some of the characters described are composites of various people and experiences I had then. It's the truth in drag.

HARPER ● PERENNIAL

P.S.™ is a trademark of HarperCollins Publishers.

HarperCollins books may be purchased for educational, business, or sales promotional use. For information please write: Special Markets Department, HarperCollins Publishers, 10 East 53rd Street, New York, NY 10022.

FIRST EDITION

Designed by Phillip Mazzone

Library of Congress Cataloging-in-Publication Data

Kilmer-Purcell, Josh.
    I am not myself these days : a memoir / Josh Kilmer-Purcell.—
1st Harper Perennial ed.
        p.   cm.
    ISBN-10: 0-06-081732-1
    ISBN-13: 978-0-06-081732-9
    1. Kilmer-Purcell, Josh.   2. Transvestites—New York (State)—
New York—Biography.   I. Title.

HQ76.98.K55A3   2006
306.77'8'09747—dc22

                                                      2005046258

        14 15 WBC/RRD 25 24 23

For Brent, who wants you to know that he had nothing to do with any events in this story. (But, I assure you, has everything to do with its happy ending.)

Let the wenches dawdle in such dress
As they are used to wear, and let the boys
Bring flowers in last month's newspapers.
Let be be finale of seem.
The only emperor is the emperor of ice-cream.

"THE EMPEROR OF ICE-CREAM"
—WALLACE STEVENS

I Am Not
Myself
These Days

Prologue

I'm freezing. The door to the balcony is wide open. The wind has blown the bedcovers completely off my feet, and the room is dark except for the faint orange glow from the skyline outside. I can't feel my toes. On the forty-second floor, the wind never stops blowing.

My boyfriend is standing over me with a knife. Two nights ago, after he had come home from a three-day crack binge, he decided that I could have the rest of the month to get my stuff together and move out of our, well *his,* penthouse. He then returned to his regularly scheduled cocaine programming and hadn't come home since. Until now.

"Why's the door open?" I ask.

"I was getting ready to kill you and then jump off the balcony," Jack says as calmly as if he were telling me what movie he was planning to see.

"With that?" I gesture toward the Wüsthof chef's knife in his hand.

"Yeah."

"But I just got that for Christmas."

It's a very expensive knife. I don't have many good things, but my parents send me one good knife each year for Christmas even though I've never used a single one for anything other than display. I think I mentioned that I wanted to take a cooking class to them once. "I haven't even used it yet."

"Sorry. It seemed like it would work best. The other ones are too small," he says, idly running his thumb over the blade.

He's right. All the other knives we have in the apartment are discount-store quality and mostly dulled from the dishwasher. They could probably do the job, but the Wüsthof would guarantee success in the first stab. Would likely go straight through and pin me to the mattress. *Wüsthof!* I imagine that's the sound it would have made as it arced through the air and then pierced my lungs, explosively releasing the air into my chest cavity. Or maybe it would have been more of a farting noise, like a balloon deflating.

"Anything holding you back now?" I ask. I doubt that he's going to go through with his plan now that I'm awake. He seems a little deflated, like I've spoiled the surprise. I'm not nervous at all. He'll do it fast if he decides to. I'm so completely exhausted after this last month we've had together that it's hard to muster up any sense of panic or impending doom.

"I just remembered which doorman is on tonight and I didn't want to scare Pedro," Jack says.

Pedro's a Puerto Rican guy in his early seventies. He's a little slow, so they put him on the late-night shift. Being a gay male escort, Jack sees him a lot on his way to and from clients. Jack likes to practice his Spanish with him.

Pedro would've been seriously upset if Jack came crashing down onto the circular drive in front of the building.

"Are we done? Can I go back to sleep?"

"Yeah. It wasn't a very elegant plan," he says. He says things like that. *Elegant plan.* This is why I still love him.

"Okay. Put the knife back in the holder. Don't leave it in the sink; the handle will rust."

"Okay." Jack starts walking back into the living room.

"Can you shut the door too, please?" I point toward the balcony. Jack's beeper goes off. Thursdays are the first busy night of the weekend. He shuts the door, grabs the cordless phone off the end table, and punches in the number on his beeper display.

"This is Aidan. You called me." Jack's voice drops an octave when he talks to a potential client. Personally I've always thought his normal voice is much sexier.

"What are you looking for?" Pause. "How long?" Pause. "Some toys are extra." Pause. "I can arrange that." Pause. "Do you need any party favors?" Pause. "Four hundred an hour."

Longer pause. "Look, you're a scum-sucking piece of shit who probably can't get it up. You're lucky I even have time to come kick the shit out of you. It's three hundred seventy-five dollars, or you can jerk off to cable porn." Pause. "I'll be over in ten minutes."

I fall back asleep while "Aidan" packs the proper dildos, restraints, lubes, toys, etc., into a backpack and heads out into the January night.

The last thing I notice before falling back asleep are the white Christmas lights strung across our balcony. From my vantage point on the bed, the little bulbs merge with the lights of the skyline behind them. They've been burning night and day for weeks and should have been taken down already. But like any chore not directly related to day-to-day survival, they'll probably stay up and lit long into spring. Things in New York City sparkle a lot longer than you'd expect before they burn out.

BOOK

I

1

I've just dropped my vodka glass and am having that perennial, silly internal debate about whether I should order another one—since, let's face it, I have reached the state where I'm dropping full glasses of vodka. A silly debate because it's highly unlikely that I will be able to keep a firm clutch on the next one, and perennial because I'm going to order one regardless. And then one after that.

I deserve another one, really. I've just broken the record for number of weeks anyone has won the Amateur Drag Queen contest at Lucky Cheng's. True, I did reuse the same song and wore the same outfit as I did on the first winning Thursday, but, honestly, this is uncharted territory here. Six weeks running of being voted the most talented amateur drag queen in New York City. By the audience. Pour me another, the future is stunning.

Of course I'm not just in it for the accolades. There's the prize money to be considered. One hundred and fifty dollars plus whatever the audience tips. After setting aside a portion for retirement, I must decide whether to invest the rest in food or two months back rent. Or possibly to retire on the spot and use it all for shots of vodka. I've retired approximately eight rounds tonight alone, not including the one that just hit the floor.

Okay, okay, already. I'll have another.

My little secret from the audience is that I'm not really an amateur drag queen. I'm practically a veteran, having been through the boot camp of drag queen training—Atlanta. Where men are men, and women are cartoon characters.

Not that I would be excluded from the Lucky Cheng's competition if my professional status were public knowledge. Quite frankly, the host of the contest I'd just won, Miss Understood, has enough difficulty rounding up three reasonably sober, mildly entertaining contestants every week. She's not going to become a stickler for rules and risk losing a weekly gig that pays her one hundred dollars and a free portion of sweet and sour pork. Besides, I've only been in New York for less than two months, so I guess I technically qualify as an amateur *New York* drag queen. Luckily, I've been able to find club work four nights a week, in addition to my day job as a junior art director at a Soho ad agency.

Miss Understood recognizes reliability, and for the moment, her name is "Aquadisiac." That's me. "Aqua" for short. Mostly just "Aqua," really. Because when I came up with the name I didn't realize that the average club-goer wouldn't catch the wordplay on "aphrodisiac." Or perhaps because it's extremely hard to pronounce with any degree of comprehension after two or three drinks. Or ten.

The name is derived from my gimmick. Every successful drag queen must have one or risk being lost in a sea of clichéd wannabes wearing Halloween novelty wigs and overstuffed bras. My gimmick happens to be fish. Goldfish usually, since they survive longest in my clear plastic tits. Not that any of them ever die in the breasts themselves. They're lovingly transferred from aquarium to tit, and tit to aquarium before and after each performance. Unless of course I happen to wake up in an unfamiliar environment, say, on a bench in Bryant Park, in which case I find the nearest faucet and refresh the tits' water supply. My mother raised me right.

I'm 6' 1" when not slouching, 7' 2" in wig and heels. My wig is blond. I wear three wigs, actually, clipped together and styled like a cross between Pamela Anderson Lee and Barbarella. My outfits are on the skimpy side: thongs, clear plastic miniskirts, vinyl boots, 22-inch corset, and a tight top with two holes cut out where the breasts should be. Into these holes slip two clear plastic domes. I purchased dozens of these clear domes from a craft store years ago. For lesser creative types than I, they were intended to be filled with holiday paraphernalia and then two of them snapped together back-to-back to form some sort of tacky oversized Christmas tree ornament. I've reengineered them with flat, mirrored backs and small holes, each plugged with a rubber stopper. They are filled with water nightly, sometimes lightly colored in honor of a holiday (for instance, tonight one's red and one's blue for the Fourth of July). The fish are slipped in through the hole in the back and the stopper is replaced. Then the tits are slipped into the evening's outfit—with small flashlights tucked underneath that shine off the mirrored backings causing the tits, and fish, to glow. All my costumes are very intricate and complicated. Marvels of modern-day engineering, really. Very often duct tape must be employed in order to keep things that mustn't be seen in places where one won't see them.

No fish has ever been harmed during an evening out. Sure, they die on a pretty regular schedule. Who doesn't? These are dime-store goldfish we're talking about. Even if I do unintentionally slash a few days off their already negligible lifespans, how many other fish can brag about meeting Leonardo DiCaprio at Limelight? Karmically, I think it's a wash.

A boy is tapping on my right breast. I tap him back on his forehead.

"If I were a petting zoo, you'd owe me five bucks. Or a drink," I say.

I have dozens of "buy me a drink" lines always on the tip of my tongue. It's imperative. I always seem to run out of retirement funds.

"They're funny. High concept," he says, still tapping on the breasts. "What're their names?"

"Left and Right. And yours?"

"Jack," he says.

"I'd shake your hand, Jack, but I have an imaginary drink in mine."

The boy is laughing. This is good. A drink is moments away. I can taste it. He's shorter than I am, though by how much, it's hard to tell. That can be determined later when I get out of the heels. He looks to be on the border of beautiful, but I need another drink to help focus. I've been fooled before. And still, I'm drinkless.

"Let's play a game, Jack of Hearts. See that long thing over there?" I ask, pointing across the crowded room. "That's a bar. The goal of this little game is for you to bring the whole thing over to me one cup at a time."

"What's the prize?" he asks.

"A free pass to the aquarium," I say, rubbing my fish tits seductively.

Finally, he heads over to the bar. Without something to focus on, my head starts to spin. I drop my head down to try to stop the mind swirl, and find myself staring directly into my tits. God, those fish are beautiful. The way they roll back and forth in the water. I think one of the flashlights has gone dead. Who cares? I've already won the goddamn prize. Fucking wig is so heavy, I can't get my head to pick up again. Maybe if I just loll it back and forth a little. Get a little momentum going.

No, that's not working. Just shaking loose a little drool. What the hell, I'll just take a little rest right here. It's well deserved. I'll need to make this Jack guy aware of my evening's triumph. He'll understand the exhaustion that comes with winning six straight weeks at Lucky Cheng's Amateur Drag Competition. Why hasn't he come back with that drink? Thomas Edison took catnaps all the time. *Where the fuck did that come from?* I think I'm talking out loud. Better shut up. Just shut up and look at the fish.

2

"Coming to" and "sobering up" are two distinctly different states of being. Each has its own independent schedule, and each comes with its own shocking revelations.

Most people aren't aware of this fact since they "sober up" sometime in the middle of the night while they're asleep, and consider "coming to" as the moment they wake up the next morning. Since my sleep averages about two and a half hours a night, I tend to approach the process from the opposite direction. For instance, on this first day into my sixth week of reign as Queen of Lucky Cheng's Amateur Drag Queen Contest, I "came to" about nine thirty a.m., and now find myself "sobering up" around one in the afternoon. I believe this reversal of the normal process owes something to the ridiculous amount of backlog my liver experiences on a daily basis. Truly, this is an organ that deserves a vacation. It's consistently voted my body's employee of the month.

"Sobering up," I find myself in the conference room at the advertising agency where I work. My day job. There are about fifteen people present—some colleagues, some clients. My copywriting partner, Laura, is sitting next to me with some storyboards that look vaguely familiar to me. I can say with relative confidence that I most likely had something to do with them. What I can't say is what they are for. The entire board room citizenry are alternating their gaze between Laura, me, and the storyboards between us.

Oh, and I'm wearing clothes I've never seen before. They're stylish and of good quality, so I don't particularly mind. I'll have to excuse myself to use the restroom later and check the tags to see who the designer is.

"With that, I turn it over to our creatives who'll present two campaigns for you."

This is Margaret speaking. Margaret is a fifty-something partner in the agency who's seen her fair share of partying herself. She has a reputation for being cold and aloof, but those of us in the booze biz just recognize that as hungover. Anyway, she's always liked me. For my birthday she gave me a collection of work by Edna St. Vincent Millay, bookmarked on her poem about burning the candle at both ends. I don't know if it was meant as a warning or as cheerleading.

People are staring at me with what I interpret as anticipation, although I'm not sure what for. Someone is going to have to step in here.

"Laura, why don't you bring us through the first campaign, and I'll take the next," I say with a smile at my fellow meeting captives, hoping against hope that there is, in fact, more than one campaign. Laura is my age, twenty-five, nearly six feet tall, full-figured and voluptuous. With her straight jet-black hair and bangs, she reminds people of Betty Page. She smiles back at me in a way that lets me know that there will be a moment in the near future when she will do me bodily harm. But for now I've bought myself a few moments to figure out what the hell is going on in this meeting, and maybe even what I'm supposed to be doing here.

"This first campaign is, basically, a testimonial campaign, about people who use Gleam-a-Lot. It opens in a bathroom with a forty-something father and his three kids all brushing their teeth . . ."

"Could one of the kids be black?" one of the clients asks predictably.

"Uh, well," Laura replies, thrown off track, "it's a family, so it would probably be simpler if they were all black. Or white."

"Well they can't *all* be *black* . . ." another client replies, looking around the table with a conspiratorial astonishment that says *"These crazy New York ad guys . . . an entirely BLACK family! How do they think up this stuff?!"*

15

"Then it's settled," repeats the original questioning client—feeling somehow vindicated, "only one of the kids is black. Please go on."

Laura's flailing, but the more she talks the more it starts coming back to me. I remember being in the agency's studio right before the meeting, with studio workers running around trying to mount Laura's and my work on blackboards. I remember sitting on the cutting bench regaling everyone with an exaggerated version of my triumphant Lucky Cheng's victory the previous night. Everyone laughs at the stories, especially the bit about the lapdance I performed on a middle-aged guy in the audience from Texas who was in New York City for an industrial textile conference. As part of my act, I got him to dial home on his cell phone, wake up his wife, and ask her how much he should tip the drag queen gyrating on his crotch.

It gets a little hazier before that, but I also remember getting into work at about eleven and finding six messages on my voicemail wondering where the hell I was, or more importantly, where the hell were the storyboards that I was supposed to have finished this morning.

Before that, things are pretty much blank. I concentrate. Blackouts can be fun if approached with the right mindset. You just can't sweat the fact that you've lost a small portion of your life for all eternity. Occasionally, little bubbles of memory will float up like surreal Mylar party balloons at unexpected times throughout the next day and start piecing together a colorful, if incomplete, version of reality.

One such balloon presently floats to the surface of my memory. It bears the caption "Pobody's Nerfect: Congratulations on Finding a Place to Sleep Last Night!" I get a flash of a bright white apartment, an amazing skyline view, and a cheese omelet with hash browns served in a round foil deli container. It's from this morning. I was at that boy's apartment. The boy from Lucky Cheng's last night. I'm not even going to attempt the mental exercise of conjuring up a name. I think I've accomplished enough. I've determined where I woke up, what meeting I'm in, and whose clothes I'm wearing. A good day's work for a Friday. Coming up with his name would just be showboating.

I can't wait for this meeting to be over and to go put my muddled head down on my desk.

I don't get home from the agency until around nine thirty p.m. The hangover that set in after lunch is finally waning. If I can keep down a soft-boiled egg and grab a half-hour nap I should be in sporting shape for tonight's midnight show at Tunnel. At least tomorrow is Saturday and I can recover a little more fully before going out again.

I've been in advertising for roughly the same amount of time I've been doing drag, which is about four years. I'm actually proud of my industriousness. Being a junior art director doesn't pay all that well, and neither does being a drag queen, so together I've calculated that I bring home roughly the same amount as a unionized sanitation worker. Unfortunately, much of the money I make as a drag queen goes back into my craft. Or down the hatch. Or, occasionally, up my nose.

There's a strange lack of knowledge about the role of drag queens in our culture. I attribute this to the appalling state of our country's education system. Others might blame an utter lack of interest. Who am I to judge?

People dressing in the opposite gender's clothing is like crack-cocaine for the daytime talk show industry. And being on television is like crack-cocaine for a drag queen. Well, actually, crack-cocaine is like crack-cocaine for a drag queen, but being on television runs a close second.

But these shows never really explain the many different types of men who dress in women's clothing. There's a big difference between transsexuals, transvestites, and drag queens. And it's this difference that I'm presently trying to explain to my mother over the phone, while boiling an egg for a light preshow dinner. A producer from Maury Povich called the other day to see if I would like to appear with my mother on an episode entitled "My Drag Queen Son Thinks I Need a Makeover!"

"Josh, I simply have no desire to go on Channel 4 with you if you plan on wearing a dress," she says.

"First of all, I'll be wearing a black latex miniskirt and halter top, not a dress. And second, it's not just Channel 4, it's *Maury*—a nationally syndicated show."

"Irregardless—" she says.

"That's not a word, Mom."

"Irregardlessly," she continues, purposely trying to piss me off, "I'm not going to have all my friends think my son wants to be a woman."

"I've explained this a million times, Ma." I sigh. "I don't want to be a woman. Transsexuals are the ones who feel trapped in someone else's body or whatever. I'm a drag queen. I'm a celebrity trapped in a normal person's body."

"Well, can't you get famous at something less perverted? Try out for a play or something. Something your father and I can go see. You were great in *L'il Abner.*"

"It was *Bye Bye Birdie,* Ma; I didn't get in *L'il Abner.* And that was middle school. Anyway, you know you're always welcome to come to my shows." I'm absentmindedly twirling the egg around in the boiling water. "And what I do is not perverted. You're thinking of transvestites,—guys who dress in their wives' clothes and jerk off."

"Language, Josh."

"Sorry, *masturbate,*" I say.

There's a long pause on the other end of the line.

"You know that if you want to have an operation that's something you can talk about with your dad and me."

"I said those are *transsexuals,* Ma. I'm very happy with my penis, thank you." I was about to add, "and so are hundreds of others," but figured I'd save it for a later conversation. My plan is to drive my mother insane with a kind of psychological Chinese drip torture, not flip her out all at once.

"Well, whatever you are, I'm not going on any cry-baby *Oprah* show with you and your weird friends."

"Jesus, Ma, are you even listening? It's Maury Povich, not Oprah Winfrey. If it were Oprah I wouldn't even give you a choice. We'll talk about it again tomorrow."

I decide that when I do get famous I'm going to have to estrange myself from my parents. I don't think they'll come across well in my *People* magazine profile.

As soon as I hang up the phone, it rings again. I'm sure it's my mother calling back to remind me to send a birthday card to my great aunt Zelda or something.

"Is Aqua there?"

"Barely," I reply.

"What's going on?"

I don't recognize the voice.

"I'm boiling an egg," I say.

"I would think a six-time Amateur Drag Queen titlist would have someone to do that for her."

"I give the staff Fridays off so they can get laid," I say.

"You're very generous," says the mystery voice.

I take a long sip from the icy vodka I'm holding in my non-egg-stirring hand.

"Well, it's easier than screwing them all myself," I reply.

I start to add up what I know about this anonymous caller.

He's someone who was either at last night's show, or someone I called today to brag about it. He's relatively humorous. And he acts as if I know him. Another good thing about being a drunk is that it sharpens my sleuthing skills.

"I was wondering if I could come by this weekend to pick up my clothes?"

Now, you'd think a question like this would narrow down the identity of this mystery caller quite considerably. In my case, it's still wide open. I have a box full of clothes from mystery dates.

"I'll bring back your drag stuff too," he continues.

"Which stuff do you have?" I ask. This'll nail it. No matter how high or drunk, I never forget an outfit and where I wore it.

"The stuff from last night. How much drag paraphernalia do you have scattered around the city, anyway?" he asks incredulously.

"Um, I was just kidding." I wasn't. But at least now I could picture him. Kinda. The boy in the mostly white apartment with the weird masks on the wall. "Can you come by tomorrow? I have a show tonight; I'll be up around noon."

"Where's the show?"

I should've kept my mouth shut. Now he's going to want me to put him on the list, and I already have two more people than I'm allowed.

"It's at Tunnel," I say, "but the list is closed."

"That's okay," he says. "It's too busy there. I wouldn't be

able to talk to you anyway. Are you going anywhere afterward?"

I think it's a little presumptuous on his part to think that I would want to talk to him anyway. I mean, sure, I went home with him, probably slept with him, ate breakfast with him, and wore his clothes to work the next day. None of this I see as necessarily flirtatious on my part. All in a night's work as far as I'm concerned. But there's something flirty/sexy about his voice that's appealing to my inner-romantic comedy actress. Then again, maybe it's just his penthouse apartment I'm hearing. My inner–gold digger frequently beats the crap out of my inner–Meg Ryan.

"Yeah, I'll probably go to the Boiler Room afterward," I say, "I should be there after three."

"Okay, I'll see you there," he says, "unless I get paged."

Paged? At three in the morning? My penthouse fantasies instantly expand into penthouse doctor fantasies. The idea of blank prescription pads just lying around an apartment nearly causes me to choke on my eggs à la vodka.

"And hey," he continues, "don't drink so much tonight. See ya."

He hangs up right before my rage begins to well up. *"Don't drink so much?!"* Who the fuck is this guy other than some guy I don't even remember fucking? No one, absolutely no one, tells me what to do when I go out. That's the whole reason I force my swollen feet into seven-inch heels that leave them covered in blisters, which would really really hurt if my feet hadn't gone numb two years ago. That's the reason I wear black latex catsuits in 100-degree dance clubs. That's the reason I spend more hours putting on wigs and makeup than I do sleeping. So that I can go to the front of every fucking velvet rope line, get showered with drink tickets and free bumps, and get paid merely to be somewhere and do whatever the hell I feel like. I have a helmet of blond hair and armor of corset to protect me from all manner of dull people—dull people who do things like watch how much they drink.

So why am I excited to see this guy?

I am, of course, completely bombed by the time I get to the Boiler Room. But being the smart gal that I am, I did a couple of bumps of coke before leaving Tunnel to appear more alert. So now I'm very alertly drunk. For instance, it's painfully clear to me that I've finished two more vodkas with no sign of this guy and I'm going to be forced to buy myself a third. The Boiler Room is a kind of after-hours hangout for the sorts of people who work in big dance clubs. So most of the other patrons have grown deaf to the constant drink requests and shrill mating calls of the North American Drag Queen.

"Hello again."

It's him. As soon as I see him, I suddenly remember the whole evening before and breakfast this morning. His name is Jack. I remember pulling into the circular drive of a contemporary high-rise apartment building on the Upper East Side and watching him hit the button marked PH. I remember walking into his apartment and thinking it was the apartment on the old *Bob Newhart Show*. Low and long and modern, with a sunken living room. The skyline of Midtown New York looked like a stage backdrop outside the floor-to-ceiling windows. I remember standing on the small balcony, with a cool summer wind drying my sweaty outfit, and looking east toward the Queensborough Bridge, which seemed close enough to reach out and touch, and much smaller than it does from the ground.

I remember him sitting on the edge of his tub watching me take off my makeup with a tube of his body lotion and some toilet paper. He watched me pull off my three pairs of eyelashes. He brought me a pitcher of lukewarm water to put my fish in. When I took off my wig, he reached over and smoothed out my flattened, messy hair with his fingers. He drew me a bath, no fussy bubbles or oils, and listened to me talk about the song I performed that night, and the audience, and my advertising job. And when I stood up in the tub, and there was no more talk, and no more Aqua, he wrapped me in a soft white towel. And I was drunk, and tired, and tired of myself, and he looked right in my eyes and said, "Hello again."

By the time he brought me the deli breakfast in bed the next morning I was performing again. Wearing an invisible wig and makeup, I was flip and dismissive and rude. Even if he never saw the clothes he lent me ever again, he was no doubt glad to see me leave his apartment.

But now, for whatever reason, he was back.

"Hey, you," I say. I try to act normal. Unfortunately I can do multiple different impressions of normal, and I can't figure out the appropriate one for the moment. I'm dangerously close to simply having to *"be myself."*

"How was Tunnel?" he asks.

"Fun. You know . . . noisy, sweaty, sceney, and sparkly."

"I haven't been in years. How was your show?" he asks.

"Good. I did a female James Bond thing. Everyone seemed to like it." I didn't tell him that halfway through the number, when I grabbed a girl out of the crowd and held her "hostage" with my toy laser gun, I was so drunk that I fell backward, pulling the mortified girl down on top of me.

"You wouldn't tell me your real name last night," he said.

"A girl's only weapon is her secrets," I say.

"Spoken like a guy with his dick tucked behind him."

"It's Josh," I say.

"Nice Jewish name."

"Actually, Episcopalian."

"I'm Catholic. Guess we'll never be married then. Shame."

He turns to the bartender and orders for us both. One Absolut cranberry, and one club soda with lime. Great. He doesn't drink. Which means one of two things: either he has never drunk, or he drank all the time and has been cut off by his "higher power." Both options are equally frightening to me, and an instant turnoff. I need to get my drag stuff back tonight, maybe have a little sex, steal a prescription pad or two, and get him the hell out of my life.

If I've learned anything, it's that sober people are just that.

"Here you go," Jack says, handing me the club soda while keeping the Absolut cranberry for himself.

I take a sip and hold the soda water up in front of my eyes, scrutinizing it.

"It's clear, like vodka . . . and has bubbles, like champagne," I ponder facetiously out loud. Holding up the lime I go on, "And this piece of fruit indicates that it's some sort of cocktail, yet . . ." I pause dramatically to take a sip before continuing, "it doesn't make my problems disappear and allow me to escape into a false world of cleverness and beauty."

"It will, however, allow you to sober up a little to better appreciate the enormous breadth of my personality," Jack says.

"Any drink that makes things look enormous is the perfect drink for me," I say, shifting my weight back on my heels.

"You don't need another drink. I'm sure you've been drinking all night," Jack says. He, of course, is right. My typical drinking schedule is one vodka rocks when I get home from work to help me nap. Two to three more while I put on my makeup and costume. And anywhere from ten to fifteen at the club. I'm not stupid. I know I drink far more than I should. Than anyone should. But part of the pattern stems from the fact that the corsets I wear bring my waist down from thirty inches to twenty-three inches. After about fifteen minutes in one, the pain becomes so unbearable that my internal organs actually begin to throb. Somehow, the booze seems to shut down my entire digestive tract and mask the pain. Coupled with the back pain and swelling feet caused by my seven-inch heels, I find that there's a level of drunkeness that I must maintain in order to finish a gig.

Jack has one more drink, and I have another soda (in addition to sipping out of someone else's abandoned rum and Coke while Jack uses the restroom) before we decide to head home. There's no question of whether and where we'll spend the night together, having already dispensed with that awkwardness the night before.

We're in a cab heading to his apartment when Jack reaches over and cups his hand over mine. It's a simple gesture, but one that catches me off guard. I keep looking out the window as we head north on FDR. Past the United Nations, the neon Pepsi sign in Queens, and Roosevelt Island. The East River sparkles under the bridges. But all I'm thinking about is his hand on mine. By this point on a ride home I'm used to a hand on a thigh or, by the more aggressive, a hand slipped up into my crotch, but not a hand on my hand.

"I don't usually go home with guys while I'm in drag," I say, breaking that silence that has held since we left the bar. I'm lying again of course. I almost never go home alone. But his hand on mine makes me want to acknowledge this trip as somehow different. If I say it out loud, maybe it's true.

"Me neither," Jack says. "But I'm not taking you home because you're in drag. I'm taking you home because I want you there."

And I believe him. I remember enough about the previous night to know that I was more comfortable at his place than I was in my own messy overstuffed studio apartment, with my horribly selfish, angry, and unemployed roommate. I know that I felt cleaner, and more content, than I'd felt in several years in his calm space in the sky. I remember the soft warm towel he wrapped around me when I stood naked in front of him in the tub.

I'm still swimming in this spa fantasy as we waltz through the lobby of his building. I often don't realize what a strange site I am to ordinary people until I notice their incredulous stares. I'm sure the doormen of his luxury Upper East Side high-rise building had never seen a seven-foot-tall drag queen with light-up fish tits clicking her high heels across their inlaid marble floors. I like Jack even more for not caring. He even calls out "hola" to one of the older doormen, who chokes out a hoarse "hola sénor" in response without ever taking his eyes off me.

While he looks for his key outside his apartment door, I sink deeper into my new fantasy life with Jack. I would wait an appropriate amount of time after I moved in before hanging up my wigs and advertising career.

As a successful doctor's boyfriend, I fully expect that I will have new duties to fulfill.

Charity would become my new hobby. Perhaps something at the United Nations, lending an international flair to my philanthropy. I picture trips to five-star hotels in third world countries after convincing Dr. Jack (I really must catch his last name) to quit his practice and join Doctors Without Borders for a year. There I'd become fast friends with caring celebrity activists like Susan Sarandon and Bette Midler and Jane Fonda, who would come over to our penthouse for fundraising brunches.

I'm coming up with the perfect brunch menu in my head when Jack opens the door and I see a middle-aged balding man lying naked and curled up on his side on the floor of Jack's foyer.

"You stupid fat pig!" Jack yells at him while giving him a sharp kick in his flabby gut. "I told you to stay in the goddamned living room, you flabby piece of shit!"

The man is hogtied. His wrists are lashed to his ankles behind his back with black leather laces. Jack grabs his wrists with one hand and his ankles with the other and drags him over to the side of the foyer so I can step in.

"I'm sorry, Aidan! I'm really sorry! I'm just a stupid fat pussy boy with a tiny, tiny dick!" the man screams as he's dragged across to the side. His sweaty clammy skin squeaks as it slides over the shiny wood floor like a new sneaker on a gym floor.

Who the hell is Aidan?

"That's right, cocksucker, now shut the fuck up," Jack says.

"I'm so bad," whimpers the man over and over.

"Quit being such a pussy!" Jack screams, "and say hello to Aqua."

"Hello," the fat man says, looking up at me.

"Don't fucking look at her you fucking piece of shit! She's too gorgeous to even notice you with your tiny useless cock!" Jack interrupts.

"Cheers," I say to the man, a little unsure if it's kosher to be friendly, or whether Jack will start yelling at me next.

"If you're a good little pussy boy, Miss Aqua will come back and let you lick her boot," Jack continues.

Jack grabs my arm and guides me toward the living room. The fat man whimpers quietly as we walk away.

"I'm sorry," Jack says, "he wasn't supposed to leave his corner." I briefly ponder what makes him think that I would be any less shocked to discover a naked hog-tied man in the living room rather than the foyer.

"*Now* can I get a drink out of you?" I ask.

Jack laughs and heads into the kitchen. I sit on the white couch and take my wig off. I can see the hog-tied man's ankles around the corner. Jack comes back with the most needed glass of vodka I've ever seen in my life.

"Umm," I start, not sure how to begin. "Who is he?"

"I call him 'Houdini.' He's a successful CEO of some huge company in London, with a wife and kids and girlfriend on the side. He comes to me one weekend a month."

"He's a patient?" I ask.

"A client. I get a lot like him. They don't even want the sex. Just want to be treated like the crappy useless weak person they feel they are inside."

"What kind of doctor are you?"

"Huh?"

"I thought you were a doctor."

"No. Who said that?" he asks.

"Never mind," I say. The drink in my hand is miraculously nearly gone. I suck at the ice. "You're a hooker."

"That's one term," Jack says. "But I don't have a lot of sex with my clients. Mostly I just beat them up."

"I'm not into that," I quickly reply, mentally plotting an escape route that doesn't involve stepping over Houdini.

"I'm not either. But it pays the bills," he says.

"More of them than wigs do, apparently," I say, looking around the immaculate penthouse. Every piece of furniture is some variant of white or light gray but is surrounded by colorful statues and masks and artwork from different regions around the globe. I see a lot of Mexican folk art, some less colorful African sculptures, some pre-Columbian masks, and some Oriental pieces scattered throughout.

"He really stays here all weekend?" I ask.

"If it's a holiday he stays three days. Sometimes four."

"What do you do with him all that time?"

"That's the best part," Jack says, smiling. "Nothing. He shows up on a Friday afternoon, I tie him up, and he spends the rest of the weekend struggling."

"How much does he pay you?"

"Two thousand dollars a day. But I usually give him a thousand back. It's good business."

"Can you get me another little sip?" I say, holding the empty glass up.

While he's in the kitchen, I walk over and peer around the corner at Houdini. He looks up at me with frightened little eyes. I wonder how old his kids are. What would they say if they saw him right now? Is this something that he asked his wife to do to him once and she refused? Does he really need to cross the Atlantic to get this kind of service? He notices me staring at him.

"Can I lick your boot?" he asks meekly.

"Um, sure." I kick my leg out toward him. At first he kisses it, with gentle little pecks. He's kind of sweet. I could see him being a good father, actually. I bet he kisses his kids goodnight like this.

After about fifteen seconds of pecking, he sticks his tongue out and licks the black vinyl with short little strokes. Soon he starts grunting and squirming to get in a better position to really go at my boot with his tongue. He's licking and rubbing his cheek against the now sopping boot like a cat rubbing against a table leg. He's really going to town. Then suddenly he bites down on my toe.

"Ow! You little fucker!" I kick him in his neck. He smiles up at me like a kid who got caught doing something bad and doesn't care.

"Do it again," he says.

"Screw you," I say, and start to walk away. He begins straining at the restraints, rolling from side to side to try to follow me. I kick him in his arm.

"More!" he yells.

"Christ, you're one fucked-up dude," I say, backing off quickly. Jack walks in from the kitchen.

"You leave her the fuck alone, pencil dick, or I'll call your wife," Jack says. Houdini stops struggling for a moment. He definitely respects Jack. Somehow I'm impressed by this.

"What do you feed him?" I ask when we're back on the white modernist sofa. I'm beginning to picture Houdini as an overweight, exotic, hairless pet cat.

"Nothing. I just put out lines of coke on the floor and a bowl of water. He doesn't get hungry. Sometimes before he goes back to the airport I'll give him a PowerBar or something."

I'm pretty much speechless by this point. I see a lot of things in the drag world. Guys who take black market hormones to grow small breasts. Trannies coming back from their operations paid for by their "patrons" and showing off their swollen new vaginas to a roomful of clubgoers. But this is strange even by my standards. I'm intrigued that there's a level of perversity even beyond my realm of expertise. Where have I been? I feel so uncool. Show me more of this party I've been missing.

I had spent so much of my time growing up being afraid of being either too cool or too uncool. Fear eventually took over and became my default emotion. If I tried to be cool, I was afraid of disappointing my teachers and parents. If I stuck with the nerd kids, I had a nagging fear that I was missing out on something. I learned to become exactly what whomever I was with at the time expected me to be. Mostly I was afraid that if I didn't become what they wanted, then they would realize what I really was. A fey little faggoty kid hiding out in a small Wisconsin town. It was an exhausting dance.

When I finally came out, the first thing I wanted to get rid of was fear. You got a problem with queers? Tough, get a load of me in this dress. You think sex is bad? Watch me tackle four guys at once. Just say no? Just say blow.

The problem with trying to be fearless is that there's always someone there to challenge your title. Like Jack. Here's a guy whom people fly across the ocean for and pay obscene amounts of money to just get screwed and have the shit beaten out of them. By comparison, I suddenly feel like the kid who plays bassoon in the junior high band all over again. Only now the cool guy likes me. I'm going to have to find a way to impress him.

"Don't you want to get out of your outfit?" Jack asks.

"Can't I just go kick him once more?"

3

hite.

White = Jack.

This was my first impression of him. And it's sticking in some fold of my brain.

The slant of morning light has slid up my sleeping body, and now pries my eyes open to the blinding flare of his bedroom.

I've woken up in enough beds that are not my own to not be unnerved by unfamiliar surroundings. I usually take a moment to say a little grateful prayer that I'm actually in a bed and not on someone's floor, or a couch in a club, or, as has happened once before, in an elevator.

Jack's bedroom is as white as the rest of his apartment. Off

to my left is a long bank of floor-to-ceiling windows interrupted exactly in the center by a glass door that leads out onto
a narrow balcony. The view from the master bedroom is the
same southern city view as the living room, except now, by
daylight, the spectacular array of buildings that make up his
backyard seem smaller, slightly farther away than they do during the night. The bed itself is white. White sheets, heavy
white comforter—even the platform that the mattress rests on
is a hard white slab made up of some sort of white marble
blocks. On the wall across from the foot of the bed is a white
television sitting on a square-edged white pedestal like the
kind one would find sculptures displayed upon in a gallery.

The only color in the room hangs on the wall above the
head of the bed. A five-foot chain of palm-sized skeletons cut
out of shiny tin and linked hand to hand like a chain of paper
men. Each skeleton is painted in a different multicolored pattern, and several are ornamented further with cutout tin top
hats, or bow ties, or twisted colorful pipe-cleaner boas, or
brightly dyed feathers arranged and glued together like an evening gown. Mexican Day of the Dead ornaments. The chorus
line of grinning garish skeletons sags across the top of the bed,
slightly fluttering in an undetectable breeze coming from the
open balcony door. When the sunlight hits one directly, a ray
of iridescent color streaks across the room and disappears as
quickly and silently as the breeze that caused it.

I'm wearing a pair of thin cotton pajama bottoms lent to
me by Jack last night. He keeps a two-foot-tall stack of them

in a drawer in the bedroom closet, all clean and fresh pressed and smelling like fabric softener. He wears a clean pair every night, he explained to me. Likes the newness. When I swing my legs over the edge of the bed, they still look perfectly pressed. I must not have turned at all in my sleep.

Impossible as it seems, the living room appears brighter then the bedroom and I squint my eyes as I pull open the door.

"Hello again," Jack says, sitting cross-legged in the middle of a fan of Sunday *New York Times* sections scattered around him on the parquet floor. It's only Saturday, but with home delivery, you get many of the fluffy Sunday sections—Real Estate, Arts and Leisure, the *Times Magazine*—a day early. Easier on the delivery guys, I suppose.

"We alone?" I ask, padding across the parquet. The vertical indentations from last night's corset still run faintly up my torso.

"Yup. Houdini had a legitimate business meeting this weekend, so he went to a hotel this morning," Jack replies. "He may be back Sunday afternoon . . . do you want breakfast?"

"I should probably go."

Jack's eyes widen almost imperceptibly, then return to normal.

"I made coffee already and ordered food before you got up. I just need to go microwave it for a sec." He's not pleading for me to stay and hang out. He's simply telling me I will. He stands up and walks toward the kitchen. "I forget what you took in your coffee yesterday," he says, rounding the corner into the dining room.

"Just a little milk," I yell after him.

In a few minutes he comes back with the deli breakfast spread and arranges the different plates in a circle on the floor.

"Sit," he says, calling me back over from the window where I was studying the skyline, "right here." He sits back down in the middle of the circle of food and *New York Times* sections and pats the floor next to him with his palm.

We spend the next hour or so eating and reading silently. When he finishes a section, he passes it over to me, and vice versa. I'm completely comfortable. It feels as easy as watching Saturday morning cartoons after a sleepover at your best friend's house.

"Listen to this," I say, breaking the silence. I point down at an article in front of me and continue. "Rioting and looting has broken out across the capital region of Sudan this morning after a grumpy overnight guest ran out of coffee and his host didn't notice." I push my empty coffee mug across the floor to him with my foot. He pretends not to notice.

"Interesting," he replies. "There's a story here about an old guy in Queens who collected every magazine and piece of mail he'd received in the last forty-five years until one of the piles fell on top of him and trapped him for five days."

He looks up at me, expectantly.

"Was he okay?" I finally reply, resigned to setting up whatever punchline was coming next.

"He was fine. First thing he did after being freed was pull himself right up and pour his own goddamn cup of coffee."

"You're a cock."

". . . And got a cup for his rescuer too," Jack continues, now pushing both my and his mug back over to me. "The rescuer reportedly likes skim milk. And a little sugar."

I roll my eyes, grab both mugs and stand up. As I turn to go to the kitchen, he leans over, reaching out to grab one of my ankles. I stop and he pulls me back toward him, causing me to hop backward on one foot, balancing the two mugs. When I'm near enough, he raises the sole of my foot to his mouth and softly kisses the very center of the arch. His lips are warm against my bare skin, and it tickles just enough to send goosebumps up my calves.

"Thank you," he says.

By the time we head out to a matinee later that afternoon I'm feeling uncharacteristically relaxed. This is so easy. *He* is so easy. I've never met anyone who just does whatever he wants to do without a thousand other little voices in his head lecturing that he's going "a bit too far," or will "pay for this in the long run," or should "just stop a minute and think twice about things." Jack doesn't seem to ever think twice. He simply has no strategy. No agenda.

It makes me wonder, what's the point of thinking twice anyway? The only possible outcome of double thinking is that you invariably end up negating whatever it was motivating you in the first place. Forcing yourself to think twice about something is just admitting that somehow you are instinctively stupid, and that repetition is the only thing that will save you from yourself.

After knowing Jack for merely forty-eight hours, I've learned that he will willingly think for me too. He'll decide when I should drink. Decide what I should wear to bed. Decide what I want for breakfast. And it's pretty relaxing, never having to think once—let alone twice—about something. The only other time I get to feel as free as this is somewhere around my twelfth vodka. It'll be interesting to see if Jack gives me a headache the next day.

4

I'm working the door at Jaguar, a small club two doors down from the Hells Angels headquarters on East Third Street. It's a slow night, and my drag codoorperson, L'il Debbie, is getting restless. L'il Debbie is close to three hundred pounds and famous for her raunchy numbers.

One of her most notorious appearances occurred just last month when she was scheduled to open a show at a club called Don Hill's. She didn't look so well when she showed up, but that was nothing terribly unusual for any of us queens.

Due to a scheduling problem with another drag queen, the hostess of the show moved L'il Debbie's song to the finale rather than the opening number. L'il Debbie sat backstage for an hour and a half, growing more and more sickly by the minute. She refused to sit down and spent the entire show pacing back and forth, sweating far more than a three-hundred-pound man in makeup and leather bustier would even under normal circumstances.

When her number finally came, she rallied. It was a high-energy punk rock version of "The Candyman." She skipped back and forth across the stage with a black leather parasol in one hand and an oversize lollipop in the other.

"WHO CAN TAKE A SUNRISE?! SPRINKLE IT WITH DEW?!" she shrieked at the audience with such force that several people actually looked as if they were trying to come up with an answer for her.

"COVER IT IN CHOCOLATE AND A MIRACLE OR TWO?! . . . THE CANDYMAN CAN!!"

She started twirling now in that way ice skaters do, where her body was in constant rotation, but her head would stop at each revolution to stare at the startled audience. She held the parasol out to her side as she spun, threatening to decapitate the entire front row.

"WHO CAN TAKE A RAINBOW?!! . . ."

The lollipop was flung out over the crowd, beaning a Long Island Italian gay boy toward the back of the room.

"THE CANDYMAN MAKES . . . EVERYTHING HE BAKES . . . SATISFYING AND DELICIOUS!!"

It was incredible that Debbie hadn't thrown up yet. She'd

been spinning at an increasing speed for nearly a full minute. The end of the song was approaching.

"THE CANDYMAN CAN!!! . . . THE CANDYMAN CAN!!! . . . THE CANDYMAN CAN!!!"

At this final line Debbie stopped spinning. What happened next will go down in drag queen history. Debbie stopped with her back to the audience, lifted up her chain-link miniskirt, and bent over, revealing what looked like a big red plastic umbrella handle coming out of her ass.

None of us could make out what it was at first. It looked familiar. All of us had seen one before. Somewhere. And then, just as the crowd collectively remembered what it was, Debbie reached behind her and started pulling and twisting at it.

It was one of those giant hollow plastic candy canes they used to sell at Christmastime in the checkout lines in Woolworths. They were about two feet long and filled with M&M's. The front row panicked as they realized exactly what that red handle Debbie was tugging at was attached to: two full feet of M&M delivery chute shoved up her ass.

The M&M's came showering out of Debbie's ass with amazing speed, bouncing off the stage platform and ricocheting into the crowd. She started swaying her hips from side to side in order to strafe the entire breadth of the audience. It was pandemonium. People screaming, then gagging after a stray M&M bounced directly from Debbie's ass into their mouth. And because she'd been delayed an hour the final few inches of M&M's had melted together and fell in a big brown clump onto the center of the stage floor.

Debbie calmly turned around to face the audience with a huge grin. She took a deep bow—extracting the tube in the process—and placed the red handle back onto it. Then she spun it around her finger and flung it out into the candy-coated shell-shocked audience.

This is a drag queen that demands your respect.

But tonight Li'l Debbie is more sedate. Neither of us has to perform. Just stand outside and let the "right" people in. Which is pretty much anyone who promises to send a drink out to us after they get inside.

"He's gotta be pretty hung to make that much money," Debbie says.

"I don't know. I haven't seen it."

"You've been dating a hooker for three weeks and you haven't slept together? What's that about? Short on cash?"

"He just doesn't want to yet. Wants us to get to know each other better first." Even as I say it I realize I sound like the naïve girl on an *Afterschool Special*.

"He's a pricey whore, and you're a cheap slut. End of story. Start fucking already," she says glibly, fishing a stray wig hair off her thigh.

Truth is, it does feel a little strange. At this moment he's on a call at a midtown hotel wearing nothing but a leather harness while beating up some naked guy he's never met. But when he stops by afterward to pick me up, we'll go to his place and sleep in pajama bottoms and modestly close the bathroom door when we pee.

"Well, like my mother always told me," Debbie says, "it's not the size that matters, it's how much it costs an hour."

"What the fuck does that have to do with anything?" I say.

"Did I say Ma made any sense, bitch?"

It's two thirty in the morning and the street is dead empty. It's one of those summer nights in New York when the bricks and pavement have soaked up so many days of relentless sun and haze that the city doesn't have a chance to cool down even in the middle of the night. If I stand next to the wall of the club I can feel the heat radiating off the building. I spray antiperspirant on my face underneath my foundation in the summer, but on nights like tonight it doesn't help. There's no such thing as a "natural look" drag queen. I'm sweaty and smeared and bored and not anywhere near drunk enough.

Three Hells Angels turn the corner onto the street from Second Avenue and roar past the front of the club on the way to their own club. The noise as they pass is incredible. I wonder how anyone can stand to live on this street.

"Hey Papa!" Li'l Debbie yells at them, waving them over. They do a U-turn in the middle of the street and come back to us. I can't help but think this might not be such a great idea.

"Hey girls," the lead motorcyclist yells back. "Club busy tonight?"

He's smiling. As are his two friends. Over the years I've learned that there are two classes of people who get a big kick out of people who are different from themselves. The very rich, and those who are freaks in their own right.

"Totally dead," Debbie yells over the roaring bikes. "Have you come to save us?"

"Get us some beer and we might do some thinking on it."

"No problem, Papa, the fridge is full," she yells back.

Debbie ducks inside and the three Angels break out their cigarettes. She comes back with bottles for them and double vodkas for us. We prop the door open so we can hear the music and start our own little block party, just the five of us. Two of them are from Queens, and one is down from New Hampshire. They show us their tattoos and pictures of their girlfriends. I show them my fish and share the secrets of "tucking" with them. Several beers and vodkas later and we're having a much better time than anyone inside the club.

Suddenly Johnny, the leader, stands up on his bike and brings his weight back down on the starter. The explosion of sound echoes down the street. He pats the seat behind him.

"Hop on, Blondie," he yells at me.

His friend starts his bike and gestures to Debbie. Neither one of us hesitate anywhere near as long as we should. I swing a seven-inch heel over the seat and settle down behind Johnny. Debbie, with considerably more effort, maneuvers her three hundred pounds onto the bike behind. With my arms around his leather waist, Johnny takes off toward Third Avenue.

The avenue is much busier, especially as we reach St. Mark's Place. I love the East Village. It should be the mandatory first home of everyone who moves to New York. We pass groups of runaway teens, half of them kicked out of their homes because they were too much trouble for their families. The other half ran away from their middle-class suburban homes intent on *becoming* too much trouble for their families. I spot Quentin Crisp in a lavender suit with matching hat and scarf trying to cross Lafayette.

Johnny is going faster now, trying to catch every green light. People stop and stare at our little group as we fly by. I'd like to think I fit in as a biker chick girlfriend. L'il Debbie, however, in her super plus–size Catholic School–girl outfit probably draws considerably more attention. I have to hold one hand on top of my head to keep my wigs from flying off, and it occurs to me that I'm not wearing a helmet.

For a second, the advanced class sixth grade hall monitor in me starts to panic. This is something I could get in big trouble for. This sort of behavior is sure to disappoint someone. My parents, my old teachers. I picture Mrs. Zariff, my fifth grade teacher, suddenly waking up with a start in her floral print bed knowing that somewhere, one of her teacher's pets is flagrantly "crossing the line." I picture my sixth grade Good Citizen Citation spontaneously bursting into flames in my old desk drawer at my parents' house. I picture my obituary in the *Oconomowoc Enterprise*, informing everyone that the brain responsible for their former debate team hero was splattered (but held loosely together by a blond wig) across Astor Place in New York City. I see teachers, priests, my parents' friends all whispering among themselves that they knew, they always knew, that I was too perfect not to have some fatal flaw. That they knew it was an act all along.

I let my eyes go blurry, finally giving in to the vodka. The streetlights and neon store signs register as streaks on my comprehension. This recklessness is just the kind of behavior expected from someone who suddenly surprises his parents with an uncharacteristic string of "unsatisfactory" marks on his third grade report card . . . the same year that he was bursting into tears inexplicably in the middle of the night. And nobody realized it was because his best friend, Greg Bransen, was moving two towns away. Nobody recognized that he had a crush on Greg, which he couldn't understand because there were no other boys who loved each other like he loved Greg in the whole wide world.

Being mutilated in a motorcycle crash is precisely the fate that good people would expect to befall a boy who moves to New York City and wears women's clothes and is developing a drinking problem and is falling in love with a guy who gets paid to have sex with other guys.

"Shut the fuck up," I tell the hall monitor in my head. *"Shut the fuck up."* What good has being good done me thus far? What have I gained other than a propensity toward panic attacks and a brief addiction to Xanax?

I'm falling in love with a hooker and it feels better than every Regional Concert Band Championship medal stuck in the back of the top drawer of the desk in my childhood bedroom. I'm falling in love with a hooker who willingly, happily, makes my life easier, and seems only to expect whatever I'm already giving him. Nothing more. Doesn't need me to bring home straight As. Or blue ribbons. Or respectable career choices.

Go faster," I yell into Johnny's ear.

5

"Maybe I could do what you do?" I say to Jack.

"No, period. Way, period," he replies between slurps of ruby red borscht.

"It says the next correspondence will be from their law firm." I'm rereading the letter I had gotten this afternoon for roughly the thirtieth time, trying to find a loophole.

"They can't kick you out for at least a year. New York City laws are all written in favor of the tenant. I know people who haven't paid rent in more than five years," Jack says.

The idea of people mooching off their landlords for more than a half decade deeply offends my inner-midwesterner. Of course my situation is different. I make money; I just don't make enough to pay my full rent. I have just barely enough cash flow to pay my half of the monthly bill, and even if I never ate or drank (shudder), I still wouldn't have enough take-home pay to cover my deadbeat roommate's half.

"Is Tempest even looking for work?" Jack asks, moving on to a plate of pierogis. We're at what's fast becoming our favorite Polish diner in my East Village neighborhood. I'm so nervous about being evicted I can't eat.

"Unless he's looking for it facedown in the laps of random cabdrivers, no," I reply. My roommate has a thing for having sex with most every penised person he comes across.

"That's inexcusable," he says.

"Not having a job or blowing cabbies?" I ask.

"Both. Because he's not getting paid for either."

When I got this job in New York five months ago, I wasn't confident enough to come to the city without a roommate. When I first saw *Miracle on 34th Street* as a child, I knew there was an apartment overlooking Central Park with my name on it. But then *Welcome Back, Kotter* got me worried that I'd have to take an elevated train covered in graffiti to get to it. Eventually, after watching *Fame,* I realized that I might just be scrappy enough to get by if I learned a heartfelt song or two and wore the right leg warmers.

But when the opportunity to move to New York finally presented itself, I froze. Atlanta was the first real city I'd lived in, moving there right after college in Michigan. And although I'd lived in Atlanta for only two years, I was pretty much Queen of the Hill, living in a little rented pink cottage in Virginia Highlands. I had my first advertising job, was doing my weekly drag show, and had a close circle of new friends. One of whom was my now-roommate, Tempest.

When I first met Tempest, he was going by the name "Piddles." Tempest had a different name every month or so. Soon after "Piddles" he became "Sarge," then "Charm," then "Grit"—each new name came with an appropriate story. One night, my group of friends and I were trapped inside our favorite club waiting out Hurricane Opal, which was raging outside. Crystal Cox, the emcee of the club's drag show, was giving periodical weather reports from the stage. She'd just finished a joke about Opal having passed through Alabama, causing "forty-five million dollars' worth of improvements" when the lights went out.

No one was quite sure what to do next. The hurricane was growing louder and more violent each passing minute. Without the club's music, we became aware of the heavy rain lashing against the outside of the building. What we thought was a heavy bass beat moments earlier was now revealed to be a near constant rolling of thunder. A huge crash startled everyone as something very heavy blew onto the roof of the club. No one had any idea what to do next. Going home, either alone or with whoever was grabbing your ass in the pitch black, was quickly ruled out as the hurricane audibly gained strength.

In the midst of this turmoil, Tempest grabbed a candle off

the bar and headed up to the stage. Most everyone in the bar had their shirts off, since the air conditioner had gone out with the power. Tempest's skin was so pale he seemed nearly translucent as he held the candle to his chest. As he passed, it seemed as if one would be able to see the flickering candle glowing right through him. His bright red hair glinted like flame itself in the strobing emergency lighting.

He took the stage, and in the faint pool of candlelight surrounding him, he began to sing an a cappella version of "Stormy Weather."

Why I thought I'd be more secure moving to New York with a man who changes names more often than his sheets is a question only five or six vodkas can answer. And after a mere four months in the city, my folly was apparent. I had yet to receive even one month's half-rent from him. Since I was the one with the stable job, it was my name that had to go on the lease. And subsequently, my name on the threatening eviction letters.

"I can give you some money," Jack said through a mouthful of pierogi.

"Thanks, but I'll figure it out. My birthday's coming up—I'll ask my parents for cash."

"Do you want to stay in my place while I'm gone?" Jack asks.

Hmmmm. Let's see. A two-bedroom penthouse with marble baths and rooftop pool? Or a shared studio apartment in a building with hallways that are permanently infused with the scent of rancid sausage and Chinese old lady pee?

"Thanks. But I'm afraid they'll move my stuff out if I'm

not there," I say. Actually, I'm afraid that Tempest will sell my things for club money. "When do you come back? Tuesday?"

"Tuesday night. Late. I'm going to give you a key. I'd love to have you home when I get there."

Jack's heading to Dallas for a long weekend at a circuit party. A certain subset of wealthy gay professionals travel around the world from circuit party to circuit party simply to get high, dance, and have sex. One of Jack's clients goes to every single one, and has hired him for the entire weekend. Fourteen thousand dollars and all expenses paid.

"I already told the doormen to let you up whenever you come by," Jack says.

I don't say anything. Maybe it's the stress of my apartment situation, or maybe it's because I'm worn out by being either drunk or hungover every day for the last several months, but suddenly I'm extremely tired and have nothing more to say.

"Hey. Don't let it get you down," Jack says, acknowledging my exhausted silence. "This is part of what it means to be a New Yorker. There's not a single person who's come to this city in the last three hundred years who hasn't spent at least one day worrying about where he was going to sleep that night. And no one's kicked you out yet." He hands me a set of his keys. Even his keys feel more luxurious than my keys.

"Thanks," I say. "Don't worry, I won't come by till Tuesday."

"Come by whenever you need to, blubberhead," he says. "That's the point of me giving them to you."

"Do you think we might be able to fuck when you get back?" I ask.

We've had this particular conversation almost every day for the last month. I haven't gone this long without sex since I came out of the closet. I'd like to say the same about him, but I'm there every night, and I hear his pager go off and see him head out to satisfy some stranger's urges. Sometimes I'm still there when he comes back and I smell someone else's smell on his skin.

Typically the conversation goes something like this:

ME: "When can we fuck?"

JACK: "I don't know. Not yet."

ME: "You get to fuck all the time."

JACK: "Are you proposing to pay me?"

ME: "Do you have AIDS?"

JACK: "No."

ME: "Something else?"

JACK: "No. I told you, I almost never even have real sex with clients."

ME: "So then it's not just me you don't want to fuck around with. You don't want to have sex with everybody."

JACK (exasperated): "I *do* want to have sex with you. I really do. I just want to wait. Like normal people."

ME: "Well, I'm normally pretty horny, so it better be soon."

JACK: "Or?"
ME: "I'll have to start sleeping with other people."
JACK: "Then we'll never sleep together."

Truth be told, he probably has the right idea. The longer we wait the more I fall for him. I can't think of anyone else who can coerce me into denying myself something I want. Here's a guy who can tell me when I've had enough to drink, and that I shouldn't have sex at the drop of a hat with him or random strangers, and for some reason, I actually listen.

It's not like I've been craving a Svengali in my life. Plenty of people have tried to get me to straighten up, sober up, whatever. It's just that at the end of the day, I don't want to end that day with the sort of people who urge others to straighten up. I want to end it with fun people. Fun people who don't want the present day to end until it's the next morning.

But waiting for Jack is, for some reason, perfectly okay with me. It feels kind of safe. A comfortable sense of inevitable gratification has been settling over me since the moment we met.

I open a packet of crackers and crumble it into the little bit of borscht he's left in his bowl. I pick up the bowl and slurp directly from it. Loudly.

I pretend to be oblivious to the obnoxious, wet sucking noises I'm making. I look up to see him staring at me.

"What?!" I ask him, mock-annoyed.

"Dickweed." He smiles at me.

6

I drop my keys onto the sidewalk for the second time. I try to laugh it off so they think I'm just clumsy and not realize that I'm so drunk and high that I can't hold on to a set of keys.

Miraculously, the keys find the hole, jiggle the exact perfect way. I try holding the door for them. The door is too heavy. I can't keep my balance on my heels and stumble back into the row of mailboxes. I cover by breaking into a quick jog to the elevator.

"Elevator's slow," I slur. "This building is crap. Not bad rent, though. The owner is on the *Village Voice*'s top ten worst landlords in New York City."

I should stop talking so much. I'm just illuminating how

drunk I am. They'll go away. They won't fuck me. I'll be alone. I'll be stupid and ugly and unfuckable.

The elevator arrives. One of the boys puts his hand out to hold the door back.

"After you," he says.

"Do you have anything to drink?" the other one asks. He's a bit shorter. Both are good-looking. They are brothers. Twins? Can't remember. I think they might have said they were twins. They don't look like twins. Similar, but not twins. Didn't one mention something about playing basketball at some college? Duke? Some place down South. Both had their shirts off on the dance floor. They came up and sandwiched me on top of the speaker I was dancing on. God, I hope Tempest isn't home. I can't wait for Jack anymore. It's too much. He doesn't even have to know. After being chaste for over a month now, I want these two like I've been in prison for years. Actually, at least in prison I would have had plenty of action. Fuck Jack. What the fuck do I owe him anyway?

"Yeah, I think I have some vodka. And maybe some scotch or rum." Thank God they want more to drink too. Now I won't have to sneak sips out of the bottle in the freezer in order to keep going. Not too much more vodka, though. I need to get up in, what? An hour and a half? What time is it now? five thirty? The gig finished at five, I talked with them for a bit, had another drink, then walked home. Maybe it's six-ish. Did I walk home? No. Couldn't have. The Tunnel was twenty blocks away. Was I in a cab with these two? Try to remember.

How did I get in the elevator? Did I press the floor button? Yes. I did. It's stopping. Sixth floor. Home.

"You got a lot of other wigs and stuff?" It was the shorter one again. Fuck off, you little faggot, I think to myself. I don't really want to spend the evening playing dress-up for these guys. I've spent the last six and a half hours in drag entertaining a room full of Long Island club trash and the very, *very* last thing I want to do is teach him how to put on makeup. I just want to drink a little more and fuck. The taller one looks at me over the shoulder of his brother and smirks. He's not going to play with lipstick and pantyhose. He wants me. Thank you. Thank you. Thank you.

"Here we are. Home. What do you want to drink?" I ask.

It's a mess inside. Every surface is covered with wigs, costumes, shoes. The only thing missing is Tempest. Thank God for small favors. Clothes from my day job are thrown in a pile on the kitchen table and chairs. In just two hours I'll need to pick out a vaguely matching pair of pants and shirt from the jumble and go to work. Jack's flight back from his circuit party job in Miami is probably just now touching down at La-Guardia. Screw him and his "I want you in my bed when I get home." For what? More cuddling? I won't be sober until after lunch. And then I'll be useless the rest of the day. Useless except that hopefully I'll have an entertaining story about the two brothers I took home and slept with the night before. Everyone at the agency will laugh at my forthrightness and lack of shame, and congratulate themselves that they are liberal and creative enough to have a drag queen to count among their acquaintances. And I will clean up the story a bit so that I was not quite so drunk, and not quite so unsafe, and I will wallow in their attentions and convince myself that they are jealous of my life, and then I will do it again tomorrow night and the next night until I am dead.

The shorter one has already picked up a dress off the floor and is slipping it over his head. It's black Lycra with strands of glittery thread sewn throughout it. He's pulling off his jeans and he looks incredibly stupid.

"You look great," I say. "Beautiful. Honestly. Here, take a wig. There's all the makeup you need in the bathroom. On the back of the toilet. It's over there." I just want to get rid of him so I can have a little time with his brother.

The taller one has gone into the living room, which is really just a small area partitioned off from the kitchen by an armoire I found on the street, which is further partitioned off into my bedroom of sorts. He starts playing with the stereo. I remember—happily—that he asked for a drink, so I fill two mismatched glasses with vodka, taking an extra swig straight from the bottle, and take them into the living room. I can barely walk and need to lean against the door frame.

"Do you have any rock?" the tall one asks.

"I don't do that," I reply. Should I offer the blow I have in the kitchen instead? No. I need that for tomorrow night. I won't have any chance to sleep until Saturday.

"Do you mind if I take off some of this costume?" I ask.

It's that weird moment. Does he want to have sex with me the drag queen, or me the boy? And do I really care one way or the other? At least he wants to have sex with me. Fuck Jack.

"Yeah, sure. Do whatever," he says.

Thank fucking God. The beautiful vodka haze blocks out most feeling, but no amount of alcohol can block the pain from the corset forever. The oppressively hot skin-tight vinyl costume is so soaked with sweat that my clammy dehydrated body chafes with every tiny movement and I thank fucking God that I can take it all off. Please let me be able to take off most everything without falling over.

So far so good. I lose my balance slightly while taking off my pantyhose, but that seems completely understandable, right? It's tricky. I balance against the armoire and take another swallow of vodka. Is this robe sexy enough? Does it ride the line between looking like me-the-guy and me-the-drag-queen he brought home? I hope that my makeup is covering any stubble grown through the night. My grateful untucked dick starts getting hard. I'm proud of this and remember the time one guy told me that even though I'd blacked out from drinking an entire bottle of vodka, I'd never lost my erection the whole time I was having sex with him and his two friends. I'm a champ. A great sport.

The taller one seems a little nervous now. No. Don't. Don't run. Don't. Leave. Me.

The shorter one comes in from the bathroom. He's wearing one of my other wigs. Its platinum shine looks completely out of place against the dark skin of his face. He makes a faggy spin to show off, and swivels his hips as he walks over to his brother. He's seducing his brother. It becomes clear from their easy familiarity with the scene that the taller brother has been having sex with the shorter brother for probably their entire postpubescent lives. I realize slowly, dully through the vodka muddle in my brain, that this is their life story. And I realize that it should be very sad, but I think that it's kind of sexy, and wrong, which makes me want to be a part of it even more. If Jack can beat the crap out of old men for money, then I can be part of a threesome with two brothers. The older one sips at his drink and peers at his brother seductively. *This is his normal.*

"Touch each other," he says very seriously, nodding at his brother and me.

And the younger one reaches over and brushes his hand across my bare chest under the robe. Suddenly, even though a second ago I was repulsed by the sight of this feminine boy, I'm now sucked into his world of wanting to do anything to please the taller, sexy brother. Even if it means taking part in some crazed pseudo–lesbian transvestite sex show.

It's hard to stand. I reach to caress him back, more than anything to keep my balance. My hand grabs his crotch through the dress. It's hard and huge and I think that maybe he's hot after all, and the room is spinning a bit and I've reached that point where between every action I take and the time I register what I'm doing is a moment of slow motion lag, and I'm not going to be able to stay upright much longer, and I lean in to kiss him to try to steady myself and my sloppy lips meet somewhere near his nose and he pushes me with all his strength into his brother and I hear the taller one scream. *"FAGGOT! YOU FUCKING FAGGOT!! DON'T KISS MY BROTHER, FAGGOT! I'M GOING TO FUCKING KILL YOU, YOU FUCKING FAGGOT!"* Why is the femme boy laughing?

CRACK.

And the light, the bright fucking light, and I am on the ground and the taller one is punching quick, sharp punches to my head. Is he hitting my face? I'm pretty sure he's hitting my face. And my head is full of pressure like when I used to swim to the bottom of the deep end of the pool and I see him—*actually see his fist hitting my eyes*—and I can't figure out why he's hitting my eyes and *FAGGOT FUCKING FAGGOT* and how can somebody be punching my eyes and yet I can still see the fist? And his knee is now pinning the side of my head to the floor, which is good because he can't quite get any more straight-on face punches, and I see the younger one grabbing my bag—my bag with the two hundred dollars I made from the club—and wigs and clothes off the floor and he turns up the stereo, why is the stereo so loud? My ear is smashed against one of the speakers and then the older one is up and kicking me in my back and I curl up with my arms over my head and he kicks the back of my head *STUPID FUCKING FAGGOT DUMB ASS FUCKING FAGGOT FUCKING*

FAGGOT FUCK YOU! And he grinds the heel of his sneaker into my ear. And they are leaving. And they are gone. And they are gone.

I need to be at work soon.

The dull gray pink light of morning is starting to slither down the eight-story airshaft and sickly ooze into my apartment. The carpet under my nose starts to heat up slightly in the sun and smells of spilt vodka and hairspray.

Jack is probably just stepping into his penthouse and realizing that it looks exactly like he left it. I'd never been there, and wasn't going to be in his bed when he walked into his bedroom like he had asked me to be. I let him down. And he let me down. He wasn't here to stop me when I needed more than anything to be stopped.

I stand up and go into the bathroom.

I look at my face.

I decide to tell people at work that I was mugged. I decide that there's not enough time between now and the time I need to be at the agency for the bruises to darken enough and the swelling to rise up enough for the public sympathy I know I will need to get through the day ahead.

I take the nail scissors on the sink and slice an even line down the side of my temple. Blood. Nobody can refute the importance of blood.

It feels so clean, the drop on my cheek. It's so much brighter than what's left of my makeup.

In my head I'm replaying what it felt like to have the boy standing over me punching me. Every time his fist connected was a relief. A puncturing of façade. A blister lanced.

I lay on the bathroom floor and I masturbate.

"Why don't you pick up your goddamn phone?!"

Laura's standing in my doorway. Her office is four doors down the hall. Apparently she's been calling me. I've been ignoring the phone, assuming it was Jack calling, wondering why I wasn't at his apartment when he got home.

"Jesus, what happened to you?" she asks.

"I got mugged," I answer.

"Idiot."

Only Laura would blame someone for getting assaulted. Then again, only Laura would correctly guess that it actually had been my own fault.

"They followed me home from the club and mugged me when I got to the door," I say.

"What'd they get?" she asks.

"My purse."

"Was your money in it?"

"Makeup, drugs, money, and several phone numbers of cute boys," I say.

"Your loss . . . cute boys' gain."

"Bitch."

My face has swollen nicely. I got into work early, since sleeping seemed anticlimactic. I had been planning my dramatic walk to the coffee machine since my arrival. I wanted to time it for the maximum size of audience. Probably about nine forty-five I figure, since advertising hours begin a little later than most workplaces. In lieu of sleep, pity would keep me going today once the drunken buzz had worn off.

I keep thinking about Jack, wondering how badly I'd messed things up with him. It was probably smarter just to ignore him. Let him call a few times, ignore him if he showed up at any of my shows, slowly let the whole thing die away. My normal process. It'd worked with dozens of guys I didn't like; why wouldn't it work with one I did? I could feel the pulsing throb in my left cheek.

"Your eye looks pretty grim," Laura notes.

"It'll go down," I say.

"You smell like scotch."

"FYI Matlock: I was drinking vodka," I say.

"Maybe the muggers were drinking scotch."

We were supposed to be concepting new ideas for Independence Life Insurance. Actually, we were supposed to be done concepting and getting ready for the presentation two days from now. We'd had the assignment for over two weeks and hadn't even sat down together once, except for lunches and the matinees we would sneak away to see. We were probably the creative team most competent at procrastinating. Our misfortune was that we also generally came up with the agency's winning idea at the last moment. The other teams had already presented their ideas internally last week and were already busy storyboarding.

"I was thinking that maybe there's an idea in their acupuncture reimbursement program," Laura begins, settling into a chair.

"Can't this wait until later?" I plead. "I'm in obvious trauma here."

"I've seen you worse."

"I need to go out at lunch and buy new makeup," I say.

"Get some that's flattering for a change," Laura says.

"Maybe you can give me some tips . . . what's that brand you wear . . . Bonne Belle?" I tease.

"Tip Number 1: Buy makeup that doesn't smear off onto someone else's fist," she says.

"Let's just get back to acupuncture."

"Okay, prick."

Laura and I have been working together since I started at the agency, and the closest we've come to saying something nice to each other is when we compliment each other's lunch order. Still, she's my closest friend outside of the club scene. She came to see one of my shows once and left in the middle. She doesn't like my alter ego. "You're an even bigger dick when you're trying to hide your real one," she said.

"Was Jack with you?" she asks.

"When?"

"When you took your makeup off with the sidewalk, loser," she says.

"No. He was out of town. I think he got back late last night," I say.

"Damn. I was hoping it was his fault."

Laura hasn't sparked to my new . . . soon to be ex . . . relationship. She met him once a couple of weeks ago when he met me for dinner after work one night. Laura can handle me coming to work drunk, my procrastination, and my insults, but she thinks dating a male escort will be the straw that breaks my brave front.

"Jack's been good for me and you know it." I was hoping she'd refute this and give me reasons to feel better about our impending breakup.

"I know you've been happier lately," she says. "But then again, I see happy people wearing cardboard coats and muttering to themselves on the street most every day."

"I'll probably see him tonight," I lie.

"Late tonight," she scolds. "We can't leave until we have one idea that doesn't make me puke."

"Let's shoot in LA this time. What's a commercial we can only shoot in LA?"

That's how we come up with our ideas. Destination first, script second. Unfortunately, no one has yet bought our "Open on a beach in Maui" stock script.

We spend most of the rest of the day thinking, interrupted

only by my frequent visits to the coffee area to regale my colleagues with the tale behind my injuries. By the end of the day, my mugging involves three youths, ski masks, a knife, and a mysterious handsome passerby who saved my life before disappearing into the black night.

Laura and I finish up around nine thirty, having two solid ideas, and one lame one involving the Eiffel Tower that, of course, could only be filmed in Paris. She has to leave to go to do whatever she goes to do at night, and I stay behind to catch up on e-mails. My phone had been ringing all day; I figured it most likely was Jack since I could see that it was an outside line. I avoided answering, telling Laura that I didn't want to interrupt our train of thought. (A train that on most days found any excuse to make multiple station stops en route.)

I linger until nearly midnight, and the office is completely empty. It's the second week of a heat wave, and I can't bear the thought of going back to my apartment. I hadn't turned the air conditioning on yet, knowing that I couldn't afford the resulting electric bill. At least I had the foresight to cancel my show tonight when I realized the swelling wasn't going down at all.

On my way out, I stop in the men's room to check out my cheek and eye. The self-inflicted cut has started to resolve itself in a perforated scab pattern. My eye grosses me out. It's full of blood and itches painfully. The other bruises and scrapes seem like they easily could be covered with foundation in time for my Wednesday-night gig.

The frigid air of the lobby shatters as I push through the revolving door out onto Hudson Street. The oppressive humidity nearly pushes me back inside. It's like a velvet curtain of heat. Immediately I begin to sweat profusely, and the wound on my cheek starts stinging.

I turn to head down King Street. It's about a thirty-minute walk home. I could take the bus, but I save the fares for heavy rain and mornings when I'm too hungover to walk.

"Hey!" a voice behind me says.

I'd told the mugging story so often today, I've nearly started to believe it myself, and begin panicking. I don't turn around. It's probably one of the mentally disabled homeless people who hang out outside the social agency next to my office building.

"Josh."

I turn around. It's Jack.

"Hey," I say. Don't engage him. Don't lead him on.

"Laura said you were still inside," he says.

We stare at each other. Laura left nearly three hours ago. And he must have been waiting for me for a couple of hours before that. He's on his rollerblades and isn't sweating at all despite the stifling heat. I try not to notice how good he looks. He's a beautiful boy. Sandy blond hair with auburn glints in the right sunlight. Or moonlight. Deep blue eyes. He's half Irish and half Northern Italian. A combination as beautiful as it is dangerous. His skin has the olive complexion of the Mediterranean region, but with the smoothness of Irish cream. I'm very in love with his lips, which are full and bee-stung and always look freshly glossed. Jack chose his work name, Aidan, when a client told him he looked exactly like a young brooding Aidan Quinn. He wears a V-neck Fruit of the Loom T-shirt almost every day, and sometimes I find myself staring at the horizontal line where the tight fabric that clings to his solid chest gives way to the soft drape that falls over his perfectly concave stomach. I like to reach underneath and rub the soft hair on his chest as I fall asleep. *Used to* like to, I remind myself.

But I'm not looking at any of that tonight. I can't afford to. It's not good for me, and I'm certainly not good for him. He's a perfectly together person with a long successful prostitution career ahead of him. I'm a wreck of a drag queen with a day job that doesn't cover my rent and a rapidly developing alcohol problem. If only he didn't have those perfect lips. And he would stop looking at me like I'm a pathetic stray street dog with his honey eyes.

We stare at each other silently, until I look at my feet.

"Come with me," he says, taking my hand.

He turns me around and we head west down Houston, taking a right on Hudson Street. He rollerblades slowly next to me. Wordlessly.

A few blocks north on Hudson, he stops at a heavy wrought-iron gate. The Church of St. Luke's in the Field. It's one of the oldest Episcopalian churches in the city and its buildings are surrounded by beautiful gardens. The largest garden is a carefully overgrown square prayer garden with ancient overhanging trees and carefully tended roses and lilies. It's a cloistered area so thick and lush that it's easy to imagine you've come across a clearing in the woods and not just wandered fifty feet away from a busy West Village street. The gate has a padlock on it.

Jack pulls on it and it magically springs open.

"If you're about to tell me you're actually the Catholic saint of padlocks, I'm going to need to sit down," I say.

He explains to me how he had come by the park before closing time and pushed the padlock together just enough so that it looked closed. When the groundskeeper came around, he assumed someone had already locked it.

We shut the gate behind us and wind through the walkways to the prayer garden. Jack sits on a stone bench and motions for me to join him.

This is when I could run away. This is when I could turn around and bolt back into the busy West Village traffic, head to a bar, get free drinks from one of my drag compatriots, pick up a cute guy, and wake up the next morning on someone else's floor with someone else's underwear wrapped around my wrists. This is where Jack and I could part ways for good, and neither of us probably would even remember the other's name by the end of the year.

I stand there. I tilt my head back and look up into the sky. I wish I could see just one star. Just one. Just one little star that would remind me of being a little kid in Wisconsin, lying with my head sticking out the door of the tent I had set up in the field behind the house. Lying there waiting for just one more shooting star, which would come at much more frequent intervals than one would imagine. In a tent that smelled of crushed field strawberries from past summers. When I thought that being twenty was being old, and that being old was no fun. And when there were billions of stars, not like tonight, when there isn't a single goddamned one, just that thick smelly orange haze that threatens to come crashing down through the overhead branches and smother me alive. I just want to see one star I saw then. A star that saw me then. So that I know there *was* a then, and not just this confusing, soul-wrenching, bruising, drunken, face-kicking *now*.

I start to cry. The tears sting the broken blood vessels in my bruised eye. Jesus Christ, I can't even cry without hurting myself.

"I'm down here," Jack says, patting the bench next to him.

I sit down next to him on the bench, less because I want to, and more because my legs are so weary I'm afraid I'll fall over any minute.

"I was mugged," I say, "I'm sorry."

"You were not mugged. Nobody gets mugged anymore in Manhattan. You brought someone home that beat the shit out of you," he says.

"I'm sorry. I'm so so sorry." I'm crying harder. It stings like a motherfucker, which makes me tear up even more.

"You are a world-class fuck-up," he says, smiling. "A class-A diamond-studded fuck-up." He gets up and starts roller-blading around the path that encircles the tree directly in front of us.

"If you weren't such an incredible overachiever in the fuck-up department," he says as he makes his first revolution, "I'd never have stood outside your office for six hours waiting for your fucked-up ass to come out the door. You fuck up where others fear to tread. You deserve an honorary Lifetime Achievement Fuck-up Oscar. And if you think you can deprive me of the joy of watching you fuck up for the rest of your life, there's one thing you need to know. I've been waiting for someone to fuck up with for a long time, and you're it, compadre."

"I don't speak Spanish," I say, lamely. I've stopped crying, but my bad eye keeps tearing.

"Further proof of your fucked-uppedness." He takes yet another loop around the tree. "And tomorrow you'll start packing up your things so that this weekend you can move in with me."

"Okay." I have nothing left inside me to argue with.

"Time for me to go, fucker-upper."

On this last trip around the tree he strips the petals of a rose blossom. He showers them over me with one hand and pulls me up with the other.

"I name this variety Summer Fuck-up. Genus: Josh; Species: Kilmer-Purcell." He skates off down the path toward the gate and turns around before reaching the sidewalk.

"See ya tomorrow, pal!" he yells back toward me, smiling, waving. Then he turns around, rolls out onto Hudson Street and disappears into the labyrinth of the West Village.

I stay on the bench, crying. For him. For me. For us. However we found each other, I know that we will never lose each other. I failed his test, and he's moved me up a grade anyway. This is not new to me. I've always gotten by on extra-credit projects. When people test me, I fail on purpose—to test *them*. We both passed, each in our own fucked-up way.

Crying always makes me have to pee, for some strange reason. I go to the corner of the Church of St. Luke's in the Field Prayer Garden and unzip myself.

It seems like the fucked-up thing to do.

BOOK

II

7

The Tempest has passed. So to speak.

It's a Saturday afternoon and Jack and I stop by my old apartment to see if Tempest has finally vacated. I gave him a week to get out after I'd moved in with Jack, and so far he's taken three. But today, other than a broken Absolut bottle in the tub and a leaking lava lamp on the kitchen floor, everything seems to have been cleared out.

I've been in New York for only six months now, and have moved from an East Village studio to an Upper East Side penthouse. I'd be lying if I didn't admit I still had a little gnawing feeling about rushing things. Then again, New York doesn't leave a lot of time for pondering forks in the road.

People who have paused to gather their wits often find themselves suddenly waking up in a cookie-cutter beige apartment in Hoboken. Or, worse yet, back in whatever backwater they came from. I will not ever leave New York. I don't know how long it takes to become a true New Yorker, but I assume that if I die here—either soonish or years from now—that that would qualify me.

Three weeks ago when I finally got all my boxes moved into Jack's, he quickly got sick of me constantly asking if "I could put something in this closet" or if he "wouldn't mind if I used half a shelf in the medicine chest." The third day after I arrived he went out on a call, telling me that by the time he returned the next morning I should have all my things put away wherever I thought they should go. "It's no longer my apartment," he clarified, "it's ours."

Of course I can't imagine anywhere I belong less. Against the stark white walls my cheap furniture and tchotchkes look like a yard sale inside the Guggenheim. But those first few days he patiently walked around the apartment picking things up and pretending to admire them. "This is a beautiful piece," he'd say about my wicker laundry hamper from Pier One. "These must be really valuable," he'd comment while admiring my collection of 1970s cereal boxes.

When he's away on calls, I feel a little more at home. Less like a hillbilly relative who's overstayed his welcome. Curiously, the only time I truly feel like I belong there is when I'm dressed as Aqua. Sometimes when I get home from one of my shows and Jack isn't there, I pour myself a double vodka in his heavy crystal rocks glasses. Still in my wig and makeup, I put on his thick white bathrobe and stroll around the apartment, looking out over the sparkling skyline pretending I am anyone from a scorned Ivana Trump to a menopausal Leona Helmsley yelling obscenities at my imaginary domestics. I make a mental note to look for heeled slippers with marabou trim on my next trip to the Chelsea Flea Market.

But by far the most amazing development in my new life is having an entire bathroom and walk-in closet just for Aqua. No more storing intricately sequined costumes in plastic grocery bags under the futon. No more ziplock bags of makeup melting in a basket next to the radiator. I have a proper dressing room and vanity. I've even bought a larger aquarium for my goldfish, and it sits on the counter in between my double sinks. I've gone from a trailer park beauty queen to Linda Evans overnight. If there ever was such a thing as a smart drag queen career move, I've just made one.

"Just leave it in there; they're going to keep my security deposit for back rent anyway," I yell at Jack, who's trying to gather the glass bits of Absolut bottle in my old bathtub. "Let's just get back to the apartment. People will be showing up in a couple of hours and I need to shower and change."

Jack's throwing me a birthday party tonight. The caterers have been at the apartment most of the afternoon cooking and setting up while we're emptying out my old home. He decided that a party would be a good way for me to meet all his friends and him to meet mine all at once. It's an interesting guest list. Most of Jack's friends are other escorts, but he still has several friends from his college school days at Columbia. He quit Columbia halfway through, and many of his friends are grad students or young professionals. I invited several people from the advertising agency as well as assorted drag queens and club kids.

Two of Jack's best friends are Ryan and Grey. They've been boyfriends for five years, and Jack has known Grey since they were in Cub Scouts together in California. When Grey first moved to New York, Jack set him up in the escort business, helping him to craft his first ad in the back of *HX Magazine,* and introducing him to the few escort agencies in New York that deal in male hookers. Grey met Ryan on call when a client had requested several boys to party with him. Ryan was one of the other escorts. He used to play the viola in a string quartet. He had terrible claustrophobia, though. His group was booked on a cruise ship and he had to quit because he couldn't even handle being on the largest ship in the fleet. Arriving in New York, he originally had dreams of becoming a classical musician. But the hooking paid much more.

The two work mainly as a team, billing themselves out as two college athletes trying to put themselves through school. Depending on the client, they either don basketball, football, or baseball jerseys before heading out on a call. One of the things I've learned about "the business" is that it doesn't really matter that Ryan and Grey are nearly thirty years old and so thin that either one would more likely be mistaken for a baseball bat than a player. Clients usually are so socially retarded and obsessed with their personal fantasies that an Asian shemale escort could answer a call wearing a backward baseball cap, rap a few bars while having sex, and leave having convinced the john that he was just fucked by a twelve-inch Rastafarian cock.

Even though Jack had told me so much about Ryan and Grey, tonight would be the first time I'd be meeting them. According to Jack, I've met them several times before as Aqua. They know which nights are mine at which clubs. They can recite lines from my shows. Apparently we've had long conversations and even shared a cab between clubs one night. I remember none of this. Even looking at pictures of them I don't recognize their faces. I'm hoping to God I haven't slept with either of them and they're simply being polite by not mentioning it to Jack.

Most of the club kids and drag queens that I've invited tonight will be meeting me out of drag for the first time. It's strange having spent night after night with these people for several months now and not having any idea what they truly look like. I wonder how many will show up in costume simply because they want to keep their "birth" persona private. Many drag queens I know even have a fictional drag history that they've concocted over the years. One claims she was abandoned at birth and raised by a secret coven of televangelists' wives. I considered being Aqua for the party, but Jack talked me out of it.

Throw in all the escorts who use fake names and it will be a party full of people who've known one another for months meeting for the first time, people who have never met meeting only half of a personality, and people who will be recognized from their escort ad as "12-inch Hot Rod" but will be introduced as Larry Feldstein. It's like a masquerade party in reverse.

Nearly every penis in the apartment tonight will at one time or another either have been tucked away or falsely advertised.

"You are the hottest guy I've ever been with," Jack says, hugging me from behind as I'm toweling off after showering for the party. He pins my arms to my side and swings me around so we're seeing each other in the bathroom mirror.

"First of all," I say, "we haven't actually *been* together yet. And second, most of the guys you find yourself *"being with"* are pot-bellied middle-aged trolls whose most attractive feature is their wallet. This is not a high bar I've managed to clear."

The party is scheduled to begin in about twenty minutes, and the caterers are taking one last smoke break on the bedroom balcony. Jack showered an hour ago and has been pacing around the apartment moving furniture around. Naked. This is yet another thing I find fascinatingly appealing about Jack. Whether it's a half dozen caterers, or the cleaning woman, or the deli delivery boy, Jack seems to feel that whomever he invites into his home deserves a free peek at the wares that pay for it. He sleeps demurely in pajama bottoms with me, but he'll greet the dry-cleaning delivery boy as if prepared to use his cock to hook the hangers on while he counts out cash for the bill.

"I mean it," Jack says, "you're the most beguiling person I know."

"That's a big word for a whore," I say.

Jack reaches around and twists my nipple.

"Ow! You realize I'm not a paying client."

He twists the other one.

"Cut it out, rent-a-cock!" I yell.

I tease Jack constantly about being an escort. Probably because secretly I find it incredibly sexy. The few people at the ad agency that I've told about his career can't believe I don't mind him having sex with other people. It's marginally more comprehensible to them when I explain that it's mostly just S&M stuff without any real sex, but I don't explain to them that I really wouldn't care if he *was* banging every single one of his clients. It's hot. I simply like the fact that I have sex with a man for free that other people pay thousands of dollars for.

Actually, I'd like it even more if we actually *were* having sex, but I fully expect the big moment to be arriving soon. What better time than my birthday?

"Why do you think I'm so hot?" I ask, looking at us in the mirror.

"Feeling a little needy tonight?" he asks.

"It's my birthday. Bring on the platitudes."

"I like you," he says, "because you're terrified."

"Can't you start out with something normal . . . like my piercing green eyes or something?"

"I'm not kidding," he continues, "I like the fact that you're terrified."

"What gives you that impression?"

Jack holds my arms tighter to my sides and bends us over the counter until our faces are inches away from the mirror.

"When I wake up every morning, I rub your arms and your chest and your stomach," Jack says, looking directly into the reflection of my eyes. "You are so relaxed, and soft, and you always smell like soap. I rub your cheek and you turn your head into my fingers. And then when you wake up, there's a moment of sheer terror in your body. Every bit of softness leaves your muscles, and you greet the day like it was the apocalypse."

"So far I don't sound the slightest bit endearing," I say, knowing in my heart that every word he's saying is true.

"You grow more electric all day. You're so stiff by the time you come home I can barely hug you."

"And then I break out the booze," I add.

"Yep. And Aqua. But instead of relaxing, you continue this self-inflicted pursuit of terror, until you finally manage to douse it with vodka and pass out. And then you're soft again."

"So far you've illustrated why we both need therapy more than why we're in love."

"I love you because you push yourself over the same cliff day after day after day and I can't believe nobody sees it but me. I love you because I've never seen anything more fascinating, and I need it to be mine. You're a shiny object." Jack pauses. "Why do you love me?"

"Have I said I do?" I ask.

"You did a second ago."

"Shit." I pause. I don't know why. I should know why. But I have no idea.

"Because you touch me when I'm sleeping," I say finally. "Because you know me soft."

The party is a bigger success than either of us imagined. Jack and I are sitting in the corner wondering if our uppity neighbors will be pissed by how large the party has grown. "Mr. Beefeater," one of Jack's regular clients, is fighting his way through the crowd to bring me a fresh vodka on a small silver tray. This qualifies as minor miracle considering the man is at least eighty years old and not terribly steady on his feet. Jack gave him his nickname based on his outfit—an authentic English Yeoman of the Guard uniform, exactly like the guy on the gin bottle.

He's been hiring Jack for several years, and his typical session begins days before his arrival when Jack gets a script in the mail outlining a scenario involving some transgression committed by a novice yeoman (Mr. Beefeater), and his subsequent punishment by the Clerk of the Cheque (Jack). By the time the old man shows up at our apartment a couple of days later with his dry-cleaning bag and hatbox, Jack has all his lines memorized. The scene can take hours to play out, with much improvised pleading and shouting. It usually involves Jack commanding the errant guard to disrobe, slowly, piece by piece.

"Off with your morion!" Jack yells during these scenes—at full volume, and with a fake English accent. *"You don't deserve the honor of that tasset! Hand over your bandolier, yeoman!"*

It's a slow motion car wreck of a peepshow, a ritualistic disrobing with increasing square inches of eighty-year-old-man flesh appearing from behind belts and medals and corsets. A drag show slowly rewinding. Jack, as the officer, remains naked throughout except for a leather harness and black boots. From what little I remember from a rushed tour of the Tower of London in high school, I doubt Jack's attire is historically accurate. But for three hundred dollars an hour, there's no limit to the revisionism Jack indulges him in. Other than the limit on the old guy's credit card.

Given Mr. Beefeater's penchant to please, Jack invited him

to come to the party tonight to be my personal servant. In full uniform and on the clock, of course. Even after living such a short time with Jack, I already consider it normal that someone is paying out hundreds of dollars an hour to serve me drinks.

Jack gets up to reward Mr. Beefeater by taking him aside and yelling at him in front of a large group of partygoers for serving my drink without a lime. The old man beams at his unanticipated punishment. Jack's generous that way.

Laura sits down in Jack's place.

"I was just in your bathroom peeing while two guys were doing lines off your sink," she says.

"Did you get lucky?" I ask.

"With the coke or the guys?"

"Knowing how you treat guys," I say, "I have to root for the coke."

"They were wondering why Jack is going out with you."

"Funny, we were having that same discussion earlier," I say. "I'm sure you stood up for me and informed them I'm plenty good enough for almost any whore."

"I said, 'Josh is the most worthy recipient of a complimentary lay of any drunk I know.'"

"Thank you, Laura," I say, "you're a true gentleman."

"Actually, I was talking to Jack earlier," she continues, "and, as much as I'm loath to admit it, I kinda like him."

"Well, he doesn't do chicks. Unless they're half of a couple."

"Seriously. He's not that bad."

"You could've taken my word for it," I say.

"If I could've understood your slurring, I might have."

The party is visually fascinating. Every drag queen I know has shown up, in all stages of costume. Any time a song with vocals comes on at least one of them hops up on a piece of furniture and does an improptu show for the crowd. As it gets later and later, I imagine that the few remaining psychologists are fighting the urge to whip out notepads and record their observations. It's like a candy factory for them. In every corner there's a different specimen. The transsexual with MS in her wheelchair in the dining room, the escort showing a group of onlookers his multiply-pierced cock in the guest bath. And in the kitchen, a small ragged gathering has circled around cooking up a batch of crack over the sink. I've never done crack, but have learned it's a favorite of the male escort industry, along with crystal meth. Unlike pot or heroin, both these drugs actually increase sexual drive, making them the drugs of choice served to the escorts paid to mingle around the better class of sex parties.

Of course, I'm getting staggeringly drunk. By the time the crowd starts to thin out, I can no longer focus long enough to realize whom I'm saying goodbye to. I make my way into the bedroom and fall on the bed. Two figures in the corner notice me struggling to remove my vinyl pants, and they team up to help the birthday boy, getting me down to my underwear and graciously throwing the sheet over me.

I'm not sure how long I've been passed out before I wake up to a warm sweaty body on top of me. It's Jack.

"Hey you," he says, smiling down at me.

"Hey sexy," I say.

"Time for your birthday spanking."

The door to the bedroom is open and I notice a few party-goers huddled in different corners of the living room. Most of the lights have been turned off, but I can see one group that's obviously in the middle of having sex. Turning my head, I see through the half-open bathroom door three people huddled over the bathroom counter lighting up their freshly cooked crack.

Jack's body starts grinding on top of me through the sheet.

"Maybe it's time I gave you your present," Jack says. Even though he's moving his hands softly over my body, I can see that every one of his muscles is perfectly stiff. And he's sweating. Hard.

He leans his face down to kiss me, and I notice his thick lips are swollen and chapped. I kiss him back. I move to kiss him again, but he suddenly gets off me and stands at the foot of the bed and rips the sheet off me.

"Come suck my cock," he says.

I'm so tired and drunk, I just want the sheet back so I can pass out again.

"I'm too tired," I slur, "just come to bed with me." A part of me can't believe I'm turning him down. I've been waiting for this for nearly two months.

"Come on. You're so beautiful. Come touch me." He grabs his dick and starts stroking it. It goes from semi-stiff to rock hard almost instantly. Even though I haven't been able to do anything with it to date, I've spent plenty of time admiring it when he walks around the apartment naked. Jack has the most disproportionately large cock I've ever seen. I've actually stood by the bed in the morning while he's still sleeping and simply stared at it. When he's lying on his back, it drapes completely over his thigh and halfway down the side of his hip. If there were a beauty contest for cocks, his would definitely carry the Swimsuit Category, and I'm hoping the Talent Portion as well.

And it's even more impressive hard.

"Come here, I want you," he says, smiling a half-smile at me, tilting his head back and laughing. It's too much for me. I crawl to the end of the bed and pull him on top of me.

I kind of expect the kind of soft-focus movie montage where we roll around making out, while the camera intercuts with close-up shots of indeterminate mingling fleshy parts. After this long a wait, I expect soft-porn movie magic moments.

But, instead, Jack's body is incredible stiff and focused. He gropes at my skin, biting me here and there, pushing and pulling my arms and legs into uncomfortable positions. He's radiating an inhuman amount of heat, and his sweat seems oily, not wet. His skin smells acrid. It tastes like he's rolled around in crushed aspirins.

We have sex for what seems like hours. Jack is so driven, I sometimes grab his head with my hands to try to force him to look at my face. When he does, his eyes don't even seem to register my presence. All the physical exertion just exacerbates my drunkenness, and I pass out and come to several times while we're fucking around. Eventually I black out completely.

When I wake up having to pee, sometime just as the sun is coming up, I'm again alone in bed. I see through the bedroom door the orgy that started hours ago in the living room is still going strong. Though some seem to have passed out at the fringes, a good half-dozen or so are still fucking like thoroughbreds on my folded-out futon. I spot Jack in the center of the group.

Mr. Beefeater has chosen to sleep on the Eames chair in the corner of the bedroom. Jack obviously hasn't let the poor man off duty. Blocks away, in a town house in Midtown where the escort agency has an office, the MasterCard machine is ticking away hundreds of dollars an hour onto the old man's card number. He hears the toilet flush, and when I come out of the bathroom, he's standing by my bed.

"Shall I get you some orange juice, master?" Mr. Beefeater asks me.

I tell him no thanks and crawl back under the covers. I try to mentally will Jack to come climb into bed with me so I can be soft and warm and relaxed. But he doesn't come.

8

It's two nights after my party and somehow I've recovered just barely enough to make a guest appearance at a club called Barracuda. The main host of the evening is a drag queen named CoCo Poof—a huge muscular African American who's famous mainly for her insults and distemper. Bar owners consider themselves lucky if an evening hosted by CoCo ends with neither her nor the audience exiting prematurely and in handcuffs.

Her show tonight is named "Snatch Game," a parody of the old 1970s gameshow *Match Game*. I'm the "celebrity" drag queen.

"Second-prize winner wins a trip backstage alone with

Aqua for ten minutes. Our first-prize winner wins a trip back-stage alone with Aqua for ten minutes, *and* a month's supply of antibiotics," CoCo announces to the bar crowd.

I use my microphone to scratch at my crotch and the audi-ence laughs. As a rule of thumb, anytime a drag queen needs a laugh, she merely needs to grab her crotch. People are fasci-nated with what a drag queen does with her penis. In reality, it's not that hard to make the thing disappear. True, some queens actually push their testicles up into the abdomen, but most, like me, merely take a hot shower and pull the whole loosened package behind them and pull on a pair of tight un-derwear. An even larger majority do nothing at all and simply wear an unrevealing outfit. I consider the latter group lazy and sloppy.

CoCo selects two guys from the front row and pulls them onstage. One is excited to get his time in the spotlight. The other looks like the kind of guy who never would normally do this sort of thing, but given the parade of empty glasses on his table, he's probably amenable to most anything at this point. Just my type.

"Aqua is so slutty . . ." CoCo says, reading off an index card.

". . . How slutty is she?!" the crowd roars back on cue.

"Aqua is so slutty," CoCo continues, "the federal govern-ment has declared her bedroom a _____."

The sound guy starts up the theme song while the contest-ants begin thinking up their answers. I scribble my answer on a card. The object is to see if either contestant matches my answer. If they do, they earn a point.

"Aqua is so slutty, the federal government has declared her bedroom a _____," CoCo repeats when the time is up. "Contestant One, I need your answer."

"Federal Disaster Area," the dramatic guy answers.

"Contestant Two? Your Answer?"

"Toxic Waste Dump."

"And Aqua, your answer?"

I flip my card over.

"Free Trade Zone."

CoCo continues her questioning. "Aqua is so drunk . . ." "Aqua is so poor . . ." "Aqua's boyfriend is so horny . . ."

I haven't seen Jack since yesterday morning after the party. When I woke up the second time, he was gone, along with the rest of the partygoers. When I opened my eyes, I was staring directly at Luis, our cleaning woman's son, who was standing next to our bed. Luis is severely autistic, and while his mother cleans, he hides in closets or behind doors and stares at us. We pretend not to see him. He gets nervous and finds another hiding place if he knows we've spotted him.

There was a note from Jack on the dining room table. Jack always covers the borders of his notes with little cryptic drawings, usually of cactuses and desert scenes. He was raised in Southern California. This note had the words "Happy Birthday" made out of saguaro cactuses across the top, and went on to say that he had a call that would last for at least a day, and that he couldn't wait to come home and crawl into bed with me. He didn't come home that night or all of today. I left some colored pencils out on the table for Luis to color in the cactuses Jack had outlined.

". . . And now it's time for our final bonus question!" CoCo shouts over the increasingly uninterested crowd. "The contestant with the correct answer to this question will take home our grand prize!" Even though the same crowd has been coming to CoCo's show for months, no one's ever pointed out to her that in addition to not actually having any real grand prize, she's also never kept any sort of score.

"Aquadisiac has a vaginal yeast infection so itchy . . ." (by now only three or four audience members are paying enough attention to slur out *"how itschy ish it?"*) "that she insists her boyfriend put _____ on his cock instead of a condom."

The DJ is nearly as bored by now as the audience is, so he plays the theme song in fast forward while the contestants and I furiously scribble out our answers.

"Contestant Number One, I need your answer! Her twat is so itchy she makes her boyfriend put _____ on his cock instead of a condom."

"A Brillo pad," shouts Contestant One, flipping his card.

"Sandpaper," mutters the shy one.

"And Aqua? Your anwer?"

"Genital warts," I answer, having completely lost all interest in anything other than a drink.

"And with five correct answers—and more important, a much shaplier ass—our first prize winner is . . . *you* . . . whatever your name is!" CoCo yells excitedly. My job now is to take the two contestants behind the curtain while CoCo performs a song. I'm supposed to get them to strip down to their underwear backstage and smear them with lipstick.

The dramatic guy begins stripping as soon as I suggest it, but the shy guy needs a little help.

"Come on," I say, "your buddy's doing it." My opening tactic is always peer pressure.

"I'll just take my shirt off," the shy guy says.

"If you go out there with your pants still on, CoCo is going to publicly humiliate you."

The shy guy stares at me pleadingly, clearly regretting that last gimlet. On to Plan B.

"Just pull them down to your knees—you can pull them up once you get out there." It's my best offer. If he declines, I'll have to start unbuttoning them myself.

He takes his shirt off and pulls his pants down to his ankles.

"That's not so bad," I say. I cup my hand over his crotch. "In fact, it's pretty impressive. I bet double your drink tickets you'll get laid tonight."

It was just the push he needed, and the pants come all the way off. I smear lipstick over both of their chests and the front of their underwear. When I'm done, I take the lipstick and write "Aqua was here" on the shy guy's back, and push them both onstage just as CoCo finishes her song.

I stumble out after them, and once I reach center stage I reach behind me and pull an opened "used" condom out of the back of my underwear, acting astonished as to how it ever could have gotten there. I hand it to the shy guy, as if returning it.

"Throw it into the audience," I whisper into his ear. He does and the audience screams and claps. Show's over. Time for a drink.

On my way to the bar I look over everyone's head as if I'm searching for someone. I've found this to be the best way to avoid people stopping me and trying to start a conversation. Not that I mind chatting with the patrons—actually that's what they pay me for—but just not when I'm on my way to the bar. Jesus Christ himself could descend directly in front of me and I would pretend to wave to someone behind him and keep beelining toward the booze.

Tucked in the corner of the bar with my sixth vodka, I put on a face like a pit bull guarding his dog bowl so that no one approaches. I scientifically calculate that I have at least four more drinks to get down before I get approximately where I want to be. No time for chitchat.

I could head home now. The bar owner is pretty cool here. Most make you stay till closing so that the bar stays full after the show, but this guy doesn't seem to care. He leaves the pay envelope behind the bar and usually heads home by one or two a.m.

Part of me wants to leave and go home to see Jack, and another part of me is petrified that he's not going to be home yet. Up until lately, I haven't minded his absences. If he's gone more than a few hours he usually calls from wherever and lets me know what's going on. Usually the long-term gigs are two- or three-day "parties" in some rich guy's hotel room. Jack calls from a phone in the lobby or outside and gives me the lowdown. How many escorts are there. How much money they're making. What kind of pervert the guy is. What sorts of drugs they're doing.

Jack always tells me what "party favors" he's done. Coke. Crack. Meth. The client insists that the escorts party with him. It's his scene. And the escorts generally make a cutoff of the party favors they supply. Jack's dealer lives on the Lower East Side, and I've seen him make it to our Upper East Side apartment in less than ten minutes. Most of the time the more professional escorts simply will pretend to use, or else find a way to take a tiny hit so they don't get too fucked up. The longer they can keep the party going, the more money they make. And if they get too high, it's harder to keep stealing the client's money. When a party's over, he and the other escorts usually head back to one of their respective apartments to sober up before heading home.

Jack hasn't called from this party.

Two hours later I'm at approximately the level of drunk I

was aiming for. Okay, so I may have gone a bit beyond. Okay, so I may have a slight problem standing. That's why God made walls.

The crowd has thinned out, but the shy guy from the contest is still here. I can't believe Jack hasn't fucking called me. We fuck for the first time, and then I haven't seen him in two days. Ass.

The shy guy keeps glancing at me, then turning away when I meet his eyes. It's cute, but a waste of my time. I wave him over.

"You have fun tonight?" I ask.

"Yeah. I don't normally do things like that."

"Who does? Where are you from?"

"I live in Jersey; I came in with my friend."

"You look like the sort of guy who likes to buy drinks for other people."

"What do you want?"

"Vodka. Rocks."

He goes off to get my drink. Jack better be there when I get home. I don't care how tired he is, I'm going to wake him up and give him shit. I didn't move in with him to be his roommate. We're supposed to be boyfriends. Real boyfriends. Boyfriends should know where the other one is every hour of every day. Boyfriends should call when their clients make them smoke crack overnight. It's common courtesy.

"Here you go." The shy guy hands me my drink. "What's your name, anyway?"

"Aqua."

"What's your real name?"

"Aquadisiac." He gives me a blank stare. This one is going to be too easy. "Are you staying with your friend tonight?"

"We'll probably head back to Jersey," he says.

"I hope you're not driving," I say.

"I'll sober up in a while."

"The bar closes in fifteen minutes," I say.

"We'll walk to a diner. Wanna come?" he asks.

"I'm probably just going home," I say. Should I head in for the kill? Logically, it's a bit premature. But I'm gonna have to go with my gut. This guy's pretty simple. "If you want, you could take a cab home with me."

"I don't know. I'm with my friend," he says warily.

"Is your friend as cute as you are?" I ask playfully.

The shy guy laughs.

"Come on," I say, "let's go. It'll be fun."

The shy guy laughs again.

"I don't usually go home with people," he says.

"You don't usually take your clothes off in front of an audience either," I reply, "but you got the hang of that pretty quickly."

He pauses and looks down at his drink. He's drunker than I thought he was.

"Why the fuck not," he says finally. "Let me go tell my friend, then we'll go."

Sometimes I wonder why I'm so predatory. I'm a monster.

A Drunk. Fucking. Monster. Now I have a boyfriend who gives me everything I want, who leaves me little notes around the house, who just threw a three-thousand-dollar birthday party for me, and I'm still a monster. Jack leaves me alone for a day and a half and I'm already prowling.

How can I bring this guy home if Jack might be there? Why am I so ready to fuck this up? So Jack is out having sex with other guys and doing drugs. I knew this about him. I can't punish him for that. He tells me everything he does like any guy would after coming home from work. "Hi, honey. Busy day at the whorehouse today. Had to beat up three men and pretend to smoke a little crack. What's for din-din?" When we're together, we're just like any couple. We read the *Times* on Sunday, we go to brunch with friends, we talk about politics and movies. He writes me cute little notes with drawings of cactuses on them for Christ's sake.

I'm not going to be a drag queen forever. And he's not going to be a whore forever. We've talked about it a little. We want to look back on this time and laugh, and tell funny old stories about our crazy days. One day we'll be real people, with stereotypical careers, clichéd midlife crises, and eventually a retirement condo in Florida.

Before the shy guy comes back, I manage to grab my purse and leave. I'm not going to mess this up.

· · ·

It's nearly four thirty in the morning when the cab pulls into the circle in front of our building. The night doorman, Pedro, sweeps me through the door with a swing of his arm.

"Buenos noches, senorita," he says.

"Buenos *días,* Pedro," I reply.

I'm not quite sure if Pedro has any idea that Aqua and Josh are the same person. But, bless his heart, he's never treated Aqua any differently than the Upper East Side matrons who glide through the lobby with their shopping bags.

I'm amazingly sober by the time I reach the forty-second floor. Sober enough to be a little frightened of what's behind our front door. I'm not sure if I want Jack to have come home and everything to be okay, or if I want him to still be missing so that I can have a reason to be angry and self-righteous.

Any questions I had melt away as soon as I open the door. Jack's standing in the kitchen —naked, of course—with the cordless phone in one hand and the deli menu in the other. He's smiling his broad goofy smile at me, and I blow him a kiss while I struggle with my heels. I hear voices from the dining room.

"Aqua!" shout Ryan and Grey simultaneously, sticking their heads around the corner.

"Hey, guys! Awfully early for you to be up for school," I reply.

I realize they're all coming down from their party. There's a sense of playful relaxation in the apartment. I imagine it's what a firehouse must feel like when all the guys come back from a four-alarm blaze.

"Make that *four* western omelets, hash browns, and wheat toast. And another large OJ," Jack says into the phone, winking at me.

I need to be in the office in less than five hours and I've had less than four hours sleep in the last forty-eight hours since my birthday party. I should simply crash in bed. But the three of them are in such a fun mood, the sky is just turning a blazing pink over the East River, Jack just ordered me breakfast, and suddenly I get my fifth wind.

"Come here, you," Jack says, coming up behind me while I'm struggling to undo my corset, and wrapping his arms around me.

I turn to kiss him.

"What happened to you, lizard lips?" I ask, noticing that his lips are more chapped than ever.

"They're just dry—the guy's apartment was over air-conditioned. Hey, I'm sorry I didn't call."

"No biggie," I lie.

"This guy was such a freak . . ." Jack starts.

"Seriously," Ryan interrupts, "I've never met a more paranoid guy. He smoked more crack than a welfare mother, and he locked the three of us in the bathroom *twice* because he thought we were undercover agents from the DEA."

"You guys go to just the *best* parties," I mock.

"And where were *you*, Miss Five A.M. Shadow?" Grey asks.

"Snatch Game at Barracuda."

"Shit, I missed it, I wanted to see you," Jack complains.

"I think I'm booked again next week."

By the time the food arrives, I've peeled off my outfit and put the goldfish back into their tank. I'm still in full makeup though, because I don't feel like showering until I eat.

Jack brings the foil deli breakfast trays into the dining room. The sun is streaming in, turning the white walls a glowing pink. Grey gets up to open the window. It's the first day of September, and a dry cool breeze fills the room. The air feels clean—like someone Windexed away the summer urine and trash smells. Like we *are* getting ready to head out to our first day of school.

We all laugh and eat as they tell me stories about their pathetic client and I tell them about the show. As the sun brightens we talk less about our night and more about other things. Jack says the sky reminds him of the Baja Peninsula, and how he used to disappear for weeks in the summer when he was in high school. Hitchhiking with his backpack down through California into the Baja Peninsula. And how his parents would simply think he'd gone to stay at a friend's house. Jack would take busses from village to village and stay with any local family that would take him in. Some nights he would just sleep by the road in the middle of the desert.

And then one day he'd just come home again and find his parents just where he'd left them, drinking old-fashioneds by the pool with their neighbors.

"Call in sick," Jack says to me.

"I can't. We have a pitch next week."

"That's next week. Come on. I want to take you to Coney Island. I haven't been all summer. We'll all go." Jack's growing more and more enthused.

"I've never been to Coney Island," I say.

"You're kidding," says Grey incredulously.

"I've only been in New York for seven months."

"Come on. Call in," Jack pleads, "I'll leave my beeper at home."

This is a serious concession. In the three months we've been together, Jack hasn't once been more than two feet from his beeper. He told me the reason he's so successful is that he's so reliable. And he is. He returns every page he gets, whether or not he's just come home from a two-day party and has only had an hour of sleep, or if we're in the middle of our entrée at our favorite French bistro. "Clients need two things—stability and unpredictability. And I know exactly when they need which," Jack's said to me before.

"Come on, go shower. Let's go," Jack says.

Before the water gets warm I have Laura on the phone figuring out how she can cover for me.

I wake up completely disoriented but also with a sense of total calm. As I come to, I feel the warm sand under my heels and the cool terry cloth under my back. I'm staring at the underside of a rainbow-hued umbrella, through which I can see faint shadows of seagulls circling overhead. I have never woken up this calm. I know I quickly will realize where I am and what I'm doing here, but for now I try to sink down deeper into this smooth haze of not knowing, not caring, not worrying.

I feel a tingling on my stomach and lift my head to find Jack drizzling a thin stream of sand through his fist.

"Wanna go to Nathan's?" he asks.

"Where are Ryan and Grey?"

"They went on some rides. C'mon, I'm starving."

Jack gets two foot-longs with onions and chili, and I just get one, with mustard and sauerkraut.

"This is going to give me major gas," I say.

"We'll have a farting contest on the way home, assuring us of our own private subway car," Jack replies, wiping chili off his chin. "You do realize you still have mascara in the corner of your eyes, right?"

"Give me a break. You gave me half an hour to get out of Aqua."

"*Hey* . . ." Jack says.

"Hey wha?" I reply, chewing a huge chunk of bun with strings of sauerkraut hanging down my lip.

"*Hey* . . . You're pretty."

"Leftover mascara smudges and all?"

"Yep."

"Well, *you're* pretty . . . *weird*," I say, dismissing him.

"Let's go to the freak show and see if we can get you an honest job."

The hot dog has done a pretty good job of placating my hangover, which has now lessened to a dull pressure in the back of my head and a slight fatigue that is somehow reassuring. I realize that I have no idea what it feels like to be sober without being hungover, so the relative lack of hangover symptoms actually makes me feel healthy. Just as the early fall weather has brushed the summer heaviness out of the air, my mind has a sense of clarity sharpened by the slight tinge of toxins in my blood.

The Coney Island Freak Show is not much of either—freakish or showy. Jack and I listen to the barker outside, promising amazing feats and stomach-churning unnatural oddities like we've never seen before.

We take our bleacher seats just as the show's beginning. Gone are the days of bearded ladies, dog-faced boys, Siamese twins, and entire tiny villages populated by midgets. I suppose there's not much heart for exploiting physical deformities anymore. Which is a shame really, since now we're stuck watching a lame parade consisting of an overly tattooed "lizard man," an old guy who pounds nails into his nostril, and a contortionist most notable for her amazing ability to inspire ennui.

"We should sell tickets to our living room," Jack leans over and whispers to me at one point. He's right. Any random weekday evening chez Jack and Aqua is more exotic than this.

"The Human Cork! Witness a man able to hold a one-liter champagne enema in his rectum!" I whisper back.

"The Chinese Grandfather Clock! Be amazed as a sixty-three-year-old Asian man suspends a swinging five-pound paperweight from his scrotum!" Jack whispers.

We spend the rest of the uninspiring show giving Jack's clients side-show names. "The Human Butt-Sniffer!" "The Insatiable Clothespin Boy!" "The Death-Defying Duke of Debilitating Dildoes!" And of course, our old favorite, simply, "High-Flying Houdini."

The sheer amount of oddness I've come to take for granted in the last seven months since I've moved to New York begins to dawn on me. As a kid, I would have to cover my eyes while watching *That's Incredible*. When John Davidson introduced a man who walked on coals, I would have to pick my feet up off the floor and tuck them safely under me. My brother could make me vomit on cue simply by turning his eyelids inside out. Do I have a growing callus over my threshold of abnormality? Or have I simply redefined normal? Maybe normal is whatever feels good where nobody gets hurt.

As we exit the freak show, Jack stops me by a Plexiglas display.

"Look, it's Aqua in fifty years."

The placard at the base of the display reads: THE FIJI MERMAID—1914. Inside, a mummified infant head is clumsily sewn onto a petrified carcass of what looks to be a salmon.

"Did this ever really fool people?" I ask, staring at the leathery corpse.

"Does Aqua?" he replies.

On the way home, the four of us are pleasantly sunburned and breezily carefree. When no one on our subway car is looking, I lick Jack's neck. His brown skin tastes warm, like the sun and the ocean. Like a freshly baked sugar cone. He smiles and snakes his hand under the bag of Russian trinkets we bought in Brighton Beach and rests it on my thigh. Jack and Grey are telling Ryan and me stories about their high school misdeeds. I never did anything remotely delinquent until well into college, so my only contribution to the conversation is laughter and admiration. In mid-sentence, Jack initiates the previously mentioned farting contest. Thankfully our subway car is nearly empty, since all-out warfare between the four of us quickly follows. The sauerkraut and chili definitely have given Jack and me an edge in the competition, but Ryan comes from behind with a surprising staccato series of stylish, almost melodic entries. Grey's contributions are less frequent, but noteworthy in their length and bass tones. None of us is willing to concede the championship title as we head, hysterically laughing, into the tunnel under the East River. For the first time I hear Jack actually giggling, uncontrollably, and I realize that I just had the best day of my life.

9

Even though we had prepared for weeks, the morning of my mother's arrival in New York registers an eight on my personal anxiety Richter scale. Like a death-row inmate, I could deal with impending doom in the abstract. But getting strapped down to the chair while knowing the executioner was on a plane speeding toward me *this very instant* is nearly enough to make me call out for a priest.

"Calm down, it's going to be cool," Jack says to me as I sit rigidly staring out over the skyline wondering about the chances of a freak airport-closing snowstorm erupting out of a clear mid-September sky. "We've got it all covered."

Ever since my mother called three weeks ago to invite herself for a long weekend, we had been concocting our game plan. Both my mother and my stepfather, who raised me and who I call "Dad," are extremely accepting of me being gay, and have been genuinely fond of my prior boyfriends. The drag thing threw them a little, but since there's little chance of their Wisconsin church friends wandering into a New York nightclub and recognizing me through three wigs and a quarter inch of foundation, they've pretty much just adopted a "don't ask, and for God's sake don't tell us about it" philosophy. I'd already cleared my schedule of all drag gigs for the weekend.

I doubted, though, that I could explain my way around a drug-dealing male escort boyfriend without seriously jeopardizing any future inheritance. Likewise, asking Jack to not wear his pager during the busy club-opening season would be overstepping the bounds of his hospitality, given that I was living rent-free in a penthouse paid for with every beep of said pager. And anyway, he had already graciously agreed not to take any in-house calls while my mother was here. Only out-calls.

It was time to get creative. We outlined several different employment scenarios that would require him to wear a beeper, and narrowed the list down to four.

1. Drug Dealer. This may not seem like an obvious mom-pleaser, but it is sufficiently close enough to the truth to explain why Jack mainly gets calls between nine p.m. and six a.m. And compared to "my son-in-law is a whore," "my son-

in-law is a drug dealer" sounds almost dignified. Still, we decide that we could come up with better, but would keep this
lie in the running in case my mother ever demanded the
truth.

2. Doctor. An obvious choice for a man and his beeper. More
specifically: proctologist. It seemed to be the one area of
anatomy that Jack could most insightfully fake knowledge of.
For a few days Jack and I practiced mock Q & A sessions
whenever he came home from a call. I'd ask him what the
proctological emergency was, and invariably he'd concoct
some scenario that would reduce us both to fits of giggling.
Plausibility was strained.

3. Nightclub Owner. Easy. It would explain the hours, and
the casual dress. After further consideration, though, we realized it would also require us to concoct a name and location.
And might even possibly mean fending off a request to physically go and visit our imaginary nightspot. Being hungover
when we came up with this idea, it was ruled out on the basis
of its simply entailing too much actual effort.

4. Travel Agent to the Super Important. Jack came up with
this one at the last minute. Living in the midst of one of the
wealthiest zipcodes in the world gave him the idea. The CEOs
and Old Rich who populated our neighborhood would need
to jetset all over the world, sometimes at a moment's notice.

117

Imagine a Park Avenue matron's frustration when she felt like an impromtu spa vaca in the Maldives with her friends and found she didn't have the right visas, or her passport had expired, or she needed to book suitable connecting flights. What would she do? Page Jack. Jack would then run all over town to different consulates, hotels, and airlines to make all the appropriate arrangements. Only a highly paid expert would be able to handle the responsibility. One rewarded amply enough to afford a penthouse apartment in the neighborhood. Naturally, many of Jack's missions would involve socialites and celebrities who would not want their comings and goings publicized, so he wouldn't be able to share too many details if my mother began asking suspicious questions. By the time Jack finished explaining his new imaginary career, my attraction to him as an escort was completely superseded by my new crush on him as an International Man of Mystery. Best of all, when I tried to explain it to my mother the night before her trip, she seemed disinterested enough not to probe.

My mother is arriving around eleven. Rather than panicking around the apartment all morning, I decide to head to work and hope for some sort of distracting ad emergency. Jack was going to meet some friend from Columbia for lunch. I left instructions with the doorman to let her in the apartment to drop off her bags, and then to put her in a taxi to the agency. Jack thought it was rude not to greet my mother immediately upon her arrival, but I explained that it was like a little welcoming present to her—I was giving her the gift of time alone in the apartment to snoop around a little. Of course we'd packed up Jack's business toys and my Aqua-phernalia and hid them safely in storage in the basement.

I spend most of the morning staring at my phone waiting for the little screen to light up with the word "receptionist."

"What's your mom's name?" Laura asks, coming in and sitting down next to my desk.

"Jackie."

"Hmmm. Jackie. Jack. How perversely oedipal."

"Yes," I reply sarcastically, "I'm dating an S&M male escort because he reminds me of my mother. They have an eerily similar spanking technique. . . . *Hey,* let's go to the Ear for a quick one before she gets here."

"It's eleven-fucking-thirty," Laura says.

"It's purely medicinal. Borderline emergency."

"Okay. I could raise a glass or two to your imminent misery."

The Ear is the local bar down the block from the agency. It's one of the oldest drinking establishments in New York, and it got its current name because the neon curves on the letter "B" in "Bar" burned out years ago, leaving only "Ear" glowing. At any time of the day odds are that someone from the agency is going to be in there. There's an unspoken rule that if you spot a colleague there, neither the spotted nor the spottee can discuss or even acknowledge it later at work. It's a more literal—and more satisfying—version of Alcoholics Anonymous.

I order a beer (because it's still morning), and Laura gets a red wine (because it's nearly lunch).

"Please, please, please come to lunch with me and my mother," I plead.

"You're so much more pathetic when you're needy," she replies.

"I'll cover for you next Friday. I'll say you're at an edit."

"Last time you covered for me you gave out three different stories," Laura says.

"The first two didn't seem convincing enough."

"If you make up a story, you have to stick with it or people realize it was all a lie."

"Maybe in Honest Land," I say. "In my world every truth is judged solely on its entertainment value."

"In your world there is no truth."

"So true. So true. Just beauty."

We spend the next two hours alternating between making fun of people at work and making bets on who's gay. By the time I realize what time it is I've had three beers on an empty stomach. Things aren't looking so dire. Except that my mother was due to arrive at the office about forty-five minutes ago. Shit.

Laura and I traipse off the elevator into our office lobby. My mother is on one of the purple sofas pretending to be interested in an old copy of *Ad Age*.

"Hey, Mom!"

"Hi! Where've you been? The receptionist didn't know where you were."

"So sorry. So, so sorry," I say, instantly regretting the number of esses I was trying to get out with my slightly drunken gay tongue. "Laura and I had a last-minute meeting uptown . . . couldn't get out of it."

"Is this Laura?" my mother asks excitedly. "I've heard so much about you."

"Hi, Jackie! Josh has been talking about you all morning!"

Laura easily pulls out a cheery midwestern persona. She was raised in Ohio.

The dread I had about Mom's visit is instantly erased the moment she hugs me a second time. She's a woman who lives pretty much only for her two sons' happiness, and it's readily apparent in her eyes every time she sees us after a long absence. It's a purity of purpose most people would be envious of. Instead of dread, I now feel guilt. Pulling one over on your mom loses its thrill sometime soon after high school. Luckily, I'm just drunk enough to get over it.

"Laura's going to have lunch with us!" I say spontaneously.

Laura shoots me a glare, but quickly recovers her midwestern charm.

"Yes, of course!" Laura replies a bit too cheerily. "Josh said he was treating us both!"

By the time we get to the restaurant Mom and Laura have completely bonded. As someone who tries to flee from any vestiges of my midwestern upbringing, I find it fascinating that Laura can sincerely alternate between New York single cynical bitch and midwestern guilelessness. Unlike me, she doesn't see it as an either/or proposition. As we take our seats at the table, she and Mom are chatting away about Laura's idea of opening a restaurant in New York City that serves only casseroles.

"So, Laura," Mom says, perusing the day's specials, "you've met Jack."

Laura does the closest thing to a spit-take that I've seen in real life.

"Sure." Pause. "Sure I have."

"I can't wait to meet him. I find what he does simply fascinating," Mom says, ruminating over the fish of the day.

This time Laura does choke on her wine. I suddenly realize I forgot to clue Laura in on the new travel agent profession we'd given Jack.

"Have you ever used his services?" Mom continues.

Laura stares at me bug-eyed. There's absolutely no way for her to answer this question.

"Umm. No. His prices are a little too rich for my blood," Laura finally replies, capping it off with a forced chuckle.

"Well, it is a beautiful apartment. He must be a busy, busy boy, running around town all day and night."

I live for moments like this. It pains me to have to interrupt.

"He'll be home this afternoon, Ma. He's so excited to meet you."

The rest of the lunch goes smoothly. My mother slowly relaxes into vacation mode. She begins chatting about the two years she lived in New York City back in the early sixties. Mom grew up in upstate New York in a small wealthy town with stuffy parents. Her father owned a cement company, which was the largest employer in town, and my father's father was the mayor. Given their positions in the village, my parents had little choice but to become high school sweethearts.

After school was finished, my mother had an inkling that maybe my father wasn't her perfect mate, but the smothering pressure to get married from both sets of my grandparents was overwhelming. In what was probably the bravest moment of my mother's life, she picked up and moved one hundred and eighty miles away to a small apartment on the Upper East Side. In retaliation, her father cut off any financial help. For two years she roomed with a nurse and an alcoholic gay poet. She became a hostess at the GE pavilion at the World's Fair.

She used to sit reading on a bench outside Beth Israel Hospital, hoping a doctor on his way home would notice her and ask her out for a drink. She wound up dating a French television producer who left her when she wouldn't give up on her dream of having children. Eventually, though, like millions of single women who have passed through New York in the last two centuries, she was defeated by the city itself, knowing that it would never offer up the marriage and family she really wanted. My future father drove into the city, packed her things in his car, and drove her home to get married, without ever proposing.

Nine years later she was divorced with two children. And shortly after, she was remarried and crammed into a puce Ford Maverick moving her kids, her new husband, a Labrador, a cat, and a hatchback stuffed with pots and pans and snowsuits to the hinterlands of Wisconsin.

Whenever I can convince my mother to talk about her short years in New York City, it's like looking through a microscope at our shared DNA. She had to get out and test her limits before returning home and building the life that she always knew she'd have.

We both were born looking for a way out of where we'd been stuck on this planet. Even if it was just a temporary reprieve. We were escape artists from day one.

And sometimes, down is the only way out.

"She's cool. I like your mother," Jack says to me in bed on the final night of Mom's visit. "She reminds me of you."

"Take that back," I say.

"She does. She says exactly the same things you say."

"She told you you have a beautiful cock?" I tease.

"Well, only once. You're much more effusive."

It had been a good weekend. A few years ago my mother and I instigated a three-day rule. We can't spend any more time together than that at one time or we descend into the level of bickering that only those who have shared one body can achieve.

Jack and Mom seemed to click. He took the three of us out to dinner each night and gave us little New York historical trivia tours on the cab rides home. Mom's a bit of a history nut, so this went over big. When I was a kid, she worked at Old World Wisconsin, one of those outdoor museums with working farms and blacksmith shops. She had to wear period costumes, and when my brother and I would spend the day with her in the summers, we would have to dress up as nineteenth-century prairie kids as well. I would work in the toolshed and lecture all the tourists on the proper method of hewing shingles on the *Schnitzlebanch*. In a way, it was my first taste of drag.

The only times during my mom's visit that were awkward were when Jack got beeped. My mother would say something silly like "no rest for the weary," or "all work, no play," and I'd feel guilty for deceiving her. By the time Jack returned he'd have concocted a convincing story about a Saudi Arabian sheik or a prime minister's son. Too convincing. It made me realize how good Jack actually was at role playing. Obviously, that's why he makes such good money, but I'd never been on the receiving end of it, and it felt weird. I love my family, and I've never been unable to be completely honest with them before. I felt something slipping away the whole time she was here.

I'm not totally sure we'd actually duped her. She puts on the face of a simple midwestern housewife, but she can be bitingly accurate in her discernment of situations. She outed me as being gay years before I finally was ready to admit it to myself.

One day when I was in high school Mom and I were in the car driving home from the mall.

"If you were planning on living a life like your Uncle Arthur, you would feel free enough to tell me, right?" she'd asked me as we merged onto the freeway.

I wanted to ask her if she meant spending my summers in a chateau in Geneva and my winters in Provence with a fabulous circle of ex-pat artists, but instead I just said, "No, of course not, where'd you get that idea?" As if I was astounded that my being president of the Drama Club and compulsive proclivity to redecorate my room with every changing season could ever *possibly* lead her to the *completely illogical conclusion that I was gay.*

"Are you sure you're not gay?" she asked me point-blank five years later when she visited me in Atlanta.

"Well," I hedged, "I might be bi."

Still, this was enough for the tears to start.

"I always knew. Always," she said.

For several months I was angry with her, and the rest of my family. If they'd always known I was gay, then why didn't anyone tell me? Take me to a few musicals or something. Send me to Uncle Arthur's for the summer. Show me that I wasn't the only freak like this in the world.

When I was ready to announce to my family that I was not bi, but a complete 100 percent Grade-A fag, I was completely underwhelmed by their response. I didn't factor in that they'd been getting used to the idea ever since I'd used Mom's Jolen Cream Bleach to highlight my hair in fifth grade.

"I'm actually going to miss her tomorrow," Jack says, turning over onto his side.

"No more hobnobbing with princes and ambassadors for you."

"Yep. Back to pig-bottoms and foot fetishes," Jack sighs.

"I, for one, can't wait to get out of flats," I reply.

"No more Normal Normans."

"It was hell while it lasted," I whisper before dropping off to sleep.

10

I am not an alcoholic. I'm a social catalyst. People pay me to illustrate for other partygoers the chemical process involved in transforming from one persona into another drunker, more fun one. It's a matter of going from dull point A to exciting point B. And I'm a raving success at it. So successful that sometimes I wind up at Mysterious Point C.

Like right now. Apparently it's morning. I've just come to, and I'm lost.

Think. Think. Think.

First the obvious. I'm Aqua, and I'm lying flat on my back across the several seats of a subway car. A subway car that's aboveground. Conclusion: I'm not in Manhattan.

Good.

Well, not *good*.

But a good first step.

Okay, the sign indicates it's an F train. This means I'm either in Brooklyn or Queens. Like a jigsaw puzzle, I'm finding the edge pieces first.

It's early in the morning. Probably Sunday. These facts are born out by the presence of a Hispanic family sitting way down at the opposite end of the car from me. They're the only other people in the train car, and the little girls are dressed in frilly lace dresses and holding white books with big gold crosses on the front of them.

They're staring at me. I suppose for good reason. When they were filing out the door this morning in their Sunday best, they probably had no idea that they would get into a subway car with a six-foot guy in a huge blond wig with fish in his breasts and stubble growing through his foundation. From my reclining position, I give them a slight finger wave with my silver elbow-length gloved hands. This ends their staring.

My head is throbbing and my stomach feels like it could rupture and send bile out of every orifice on my body—including my navel. This means I've probably been drinking heavily at some point in the not too distant past. Of course that could just as easily have been concluded from the "waking up lying down across subway seats" portion of the morning.

I give myself a once-over, a feat easily accomplished by raising my head slightly off the hard orange plastic seat. I'm wearing my Jane Jetson outfit—a silver stretch vinyl top, turquoise water for the fish, and a silver miniskirt that's currently hiked up enough to generously show off my matching silver thong. A quick check on the fish shows them happily swimming around in the morning sun.

And one of my silver glitter seven-inch platform shoes is missing. Hmmm. I'm going to put that mystery aside for the moment. Perhaps after I've had some coffee.

Okay. No blood or bruising. Happy Happy Sunday. *See, Hispanic Bible Kids, God loves drag queens too.*

Now, I'm not going to panic because my bag with my money and ID is bound to be tucked away safely under the seat.

Only it's not under the seat.

Thankfully I still have a little buzz going or this could really be disconcerting.

Silver lining: I'm already on the subway so I don't *really* need any cash anyway. I'll simply get off at the next stop and get on a train heading back into town.

Some people might get obsessed with figuring out how they wound up on the F train in drag, with no bag and only one shoe, but that's simply not my style. What's done is done. I'm sure I had my reasons.

I pull myself up with a little help from the pole and wait for the train to pull into the next stop. The Hispanic family is leering again. As the train pulls into the station I stand and hobble over to the door. Before I exit, I turn to the family and bestow a papal blessing on them with an outstretched gloved arm.

Dignity is in short supply as I board the Manhattan-bound train back. Thankfully there's a bedraggled homeless person in the car. I sit as close to him as I can stand, given the smell, hoping that when people stare at me, they might at least think, "Well, at least she's better off than the homeless guy next to her."

Two hours, three transfers, and one pit stop to vomit into a trash can later, I'm hobbling down Second Avenue toward home. It's actually a beautiful morning. When my mother left a week ago it was unseasonably cold and rainy, but this morning is bright and clear and warm. This is fortunate since I barely have any clothes on. Will my luck ever run out?

I may have to admit to myself that I went a little bit overboard this weekend. Perhaps I had a little bit of Aqua backlog in me after my mother's visit and I had to get it out of my system. But it's a lovely Indian summer morning, I have all day to recover, and I'm sure Jack's at home waiting for me with a fresh pot of coffee.

I pick up the *New York Times* outside our door and ring our bell, since my keys are in my missing bag.

"Who is it?" comes an unfamiliar hoarse voice from the other side. I double-check the apartment number in case I've gotten off on the wrong floor.

"It's me. Aqua. Lemme in," I shout back.

There's a rustling inside, and as it gets closer there's a sound like somebody throwing themselves against the door. More rustling, and what sounds like metal scratching against the door, and finally after much effort the deadbolt clicks open. A thud as whoever it is slides down and hits the floor, and then, "Come on in."

I turn the knob and open the door. Houdini is lying on the floor of the foyer, naked, handcuffed and hogtied in his usual manner.

"Hey. How ya doin. Thanks for getting the door," I say.

"No problem," he says, as though using one's teeth to manipulate a deadbolt while one's hands and feet are tied behind one's back didn't require a superhuman level of concerted effort.

I step over him and put the paper on the kitchen counter. No coffee. Shit.

"Aidan's not here?" I ask, remembering to use Jack's work name.

"He had another call," Houdini replies.

"You're an awfully generous man to let him date other people," I tell him.

"I needed a break. Hey, can you cut some lines down here for me?"

It's the least I can do for the guy, given what he went through to get the door open. I grab the packet and blade on the kitchen counter and divide three lines for him on the parquet floor.

"That enough for you?" I ask.

"For now, thanks."

"Do you want me to untie you?"

"NO! Aidan would kill me!"

Wishful thinking, I think to myself while I get a bowl of fresh water for the guy. I head into my bathroom to begin the hour-long chore of getting out of Aqua.

I'm starving by the time I'm through. All I can think about is a greasy bacon, egg, and cheese roll from the deli downstairs. Except that my missing bag had all the cash I had to my name.

"Hey, do you by any chance have a few bucks you could lend me? I'll pay you back when Aidan gets home," I ask Houdini.

"Sure. My wallet's in my pants. I think Aidan put them in the hall closet. It's mostly pounds, but I think I have a few American dollars in there. Help yourself," Houdini replies.

I can't believe I'm taking money from a guy tied up on my floor. It's like an absurdist crime scene. Strange man enters home, gets tied up, resident leaves, strange man opens door for roommate, roommate steals all his money.

"Do you want anything?"

"No, I'm fine. Thanks." The coke is doing its job. Houdini follows my movements with anxious flitting eyes.

I call in my breakfast order and grab the paper before heading into the living room. I'm exhausted, but I know if I go to sleep now I'll wake up this evening and not be able to sleep all night. I'm halfway through the Styles section when I consider calling back the deli to add tomato juice to my order so I can make myself a Bloody Mary. It'll help smooth out the day a little.

When breakfast comes, I can only open the door a few inches so the delivery guy won't see the tied-up naked guy in our foyer.

"Sure you don't want any?" I ask Houdini again, the polite midwestern hostess in me taking over.

"Nah, I'm fine," he replies. "Can you help me get into the living room, though? I'd like to lie on the futon a bit. I'm getting cramps."

I help him get semi-upright and he crawls on his knees into the living room. I grab a plate and follow him. He keels over on the cushion next to my chair.

"Want some of the paper?" I ask.

"I'm fine. I'll just wait for Aidan to get back."

"How long are you here for?"

"Just a day and a half. I go back late tonight. On the red-eye," he answers.

"What do you do over there? Do you mind my asking?"

"I'm CEO of a specialty foods distributor," he answers.

"Aidan told me you're married."

"Twenty-two years. Three kids, a daughter, and two sons."

"Nice," I say, not sure how to continue a casual conversation with a sky-high, bound-up, naked CEO. I go back to my paper.

"What do *you* do?" Houdini asks me a little later.

"Well, I'm a drag queen at night and an advertising art director by day," I reply.

"That sounds fun."

"Which?" I ask.

"Both," he replies.

"Not really," I sigh.

"Which 'not really'?"

"Neither."

"Neither?"

"N-I-ther."

"Let's call the whole thing off." So Houdini's a funnyman. I never would have guessed.

"How long have you been dating Aidan?" he asks.

I add it up in my head.

"Only about three months," I say.

"And you don't mind what he does for a living?"

"Does your wife mind what you do for fun?"

"She doesn't know about it," he says.

"Well, I'm one up on you there."

"Aidan's really good at what he does," Houdini says. "I've tried a lot of people. Masters, Mistresses. All around the world. He's the best."

This makes me proud. The best dominating, humiliating, physically abusive whore in the world is mine. All mine. Almost makes me wish I were into getting beat up, just to take full advantage of his talents.

"What makes him so good?" I ask.

"I dunno. He just never lets up. Sometimes even when I'm at home I feel like he still owns me. It's just a residue in my mind all the time. Like I can't fuck up or he'll know."

"I know the feeling," I say, and then pushing harder, "Why do you do this anyway?"

"I don't know. I really don't. It's just a different place in my head. Where everyone isn't saying 'yes' to me all the time," he says. "And I like the drugs."

"Fair enough."

"Why do you dress in drag?" he asks me.

"It's just a different place in my head . . . and I like the free booze."

"Maybe I'll take a section of that paper now," Houdini says, "only you'll have to promise to take it away as soon as Aidan gets back."

"Deal." I spread the Business section out on the coffee table in front of him so that he can turn the pages with his teeth.

Houdini and I spend the rest of the morning this way, as if I had an old friend over for brunch. I trade out sections of the Sunday paper for him as he's ready, and we read each other interesting bits of articles we're reading. The Saatchi gallery in London has an exhibit we find interesting, and he invites me to visit there with him if I ever find myself in London.

Occasionally, while I'm reading something I catch him out of the corner of my eye struggling against the wrist restraints behind his back. I don't think he's even aware he's doing it. Just a subtle straining and rhythmic twisting of his forearms against the leather. I'm a little jealous of him. Such a straightforward manifestation of his subconscious. Embracing his trap. Knowing what he's fighting against. People spend thousands of dollars on psychologists and doctors every year of their entire lives just trying to find out what's holding them back while their mysterious enemy slowly decimates their day-to-day lives. Houdini shells out a few grand every couple of months, confronts his demon, and heads back to his successful life stronger than before. He makes it seem so simple—on a par with exercising regularly and eating right.

I give in and have a small vodka shortly after noon. Lying on the couch in my underwear, the cool breeze from the window skips across my skin and I close my eyes and listen to the muted traffic noises far below. Houdini's still wide awake from the three more lines of coke I cut him between the Week in Review and the Travel sections. Without any pen or free hands to hold one, he's silently pondering the crossword puzzle as I slip softly into a deep nap, thinking of the city, and its spaces, and the lulling waves of Sunday happening all around me. For a split second before I fall asleep I realize I'm totally relaxed.

11

This is how I become not me:

It is an exacting process—there's no room for error, and little for improvisation. It is ritual and sacred, and regardless of my physical or mental condition, it is unchanging.

It begins by monitoring my diet for the entire day before any show. My body must be relatively empty of food to fit into the corset, and relatively full of alcohol to dull the discomfort.

About four hours before I head out, I gather together the pieces of my predetermined outfit.

Two pairs of pantyhose. Up to three wigs—combined together and prestyled. Tucking panties. Decorative panties or thong. Matching elbow-length gloves. Bag. Shoes. Necklace. Earrings. Assorted accessories. Wig cap. Toys—laser guns, bubble makers, candy to toss out into the crowd. All is transferred, piece by piece, into the bathroom.

No one is allowed to witness the transformation. It occurs completely behind the closed bathroom door. It's a slow motion magic act where the male audience volunteer disappears into a box and a woman appears from inside hours later.

Before sequestering (quarantining?) myself, the final ingredient is procured from the kitchen. Two large glasses of ice and the bottomless liter bottle of Absolut that lives in the freezer. The glasses are carried in one hand, the bottle in the other, and I disappear into the bathroom completely naked.

Inside, I pour my first glass and lean into the mirror to inspect my face. Is my stubble long enough? If I didn't have a business meeting earlier in the day, I don't shave. The longer the whiskers, the cleaner the shave will be. Are there any particularly prominent zits that will need extra cover-up attention? My eyes are relatively deep set and heavy lidded, so I must keep my brows plucked as thin and high as they can be to create maximum real estate for the dramatic application of multiple shades of eye shadow.

After I've concocted my facial plan of attack, I sit on the toilet. Whatever I can get out of me will mean less pain in the corset later on. Also, while it's possible to wriggle myself out of an outfit just enough for an emergency piss, any other kind of bathroom maneuver would require a near complete dismantling.

By now I've finished one tall glass of vodka and am ready for the shower.

The water must be scalding hot and the air in the bathroom as humid as possible. I shave my entire body each night I have a show. A rigorous shaving schedule, religiously adhered to, reduces the outcropping of ingrown hairs. I start from the bottom up. I'm lucky enough not to be terribly hirsute, otherwise I would need to add another hour to my prep time.

Toes first. Then legs. Then genitals and ass. I move up to my navel, and then remove the few scattered chest hairs I'd prayed so ferverently for when I was entering puberty. A double shave under the arms. Then, when I reach my neck, I start over again at the toes and give my body a second smoothing. Nothing escapes my razor.

Blade change. Gulp from vodka resting on edge of tub.

On to the face. Knowing that I could be out anywhere from four to twenty-four hours requires the closest shave I can deliver. Two shaves with the grain, two shaves against. Complete removal of sideburns, and about half an inch from the perimeter of the entire hairline. Even a peek of brown hair poking out from under the edge of the wig will completely ruin the illusion.

A quick shampoo, all-over soaping, and I'm out.

The quickest but most vital part of the transformation is

next. While my body's still steaming, and my twig 'n' berries are at their most relaxed, I pick up the pair of nude-colored, two-sizes-too-small, spandex panties off the floor. I slip both feet through the leg holes and pull them up to my knees. Then I spread my legs slightly, bend deeply at the waist, and reach around behind me with one hand. Twisting my hand between my legs from behind, it would look to some as if I were trying to sneak up on my own unsuspecting genitalia. Or practicing an obscure yoga pose. "Downward Facing Python Chase." Grasping my surprised triumvirate in one hand, I pull the whole package backward as I yank up the tight panties with the other. Trapped. Straightening up from the bend, I can feel my lower abdomen stretch and flatten as my precious goods give up and settle into their new hideaway.

At this point, I sit on the tub edge and sip more vodka while I wait for the steam to clear from the bathroom. I watch myself appearing slowly through the fog on the mirror. Without the body hair and visible genitalia an apparition of an androgynous mannequin sits staring back at me. Sometimes, if I'm particularly preoccupied with the prior day's events I'll switch the lights off at this point, trying to switch my mind from Josh to Aqua. It can happen in a moment when I'm not concentrating on it, or it can take the entire preparation time and not be completed until the wig settles into place. It's entirely unpredictable, and as many times as I've undergone the transformation, I have no concrete mental process to force the handover.

Then the shift from destruction to creation.

I slick my short wet hair back and snap on the nylon wig cap. The boring parts get foundation first. Neck, ears, "décolletage." I only use MAC products. Only. I'm not sure if it stems from my advertising background, but I'm completely brand-loyal.

In the summer, I roll or spray a very light layer of antiperspirant onto my face. Then with a slightly darker shade of foundation, I sponge contouring shadows on either side of my nose to make it appear thinner, and streaks on my temples and cheeks to give the illusion of delicate bone structure. By now I look like a splotchy paint-by-number portrait.

Next, the overall foundation application takes place. With a clean sponge, I smoothly stroke the thick fleshy putty over my face and neck, carefully blending the edges around my ears and neck. Then the powder. As I press it into the moist foundation, taking care to fill the creases around my nose and eyes, my skin takes on the completely even texture and color of a blow-up doll. It takes a delicate measuring hand—too much powder and cracks will develop when I laugh; too little and I'll look like a greasy circus clown

On to the fun part. Eye shadow and lipstick. Depending on the outfit, I choose either a bold color palette or simple matching earth tones. I lay the brushes and sponges out on the counter like surgical tools. Often I need to mix pigments and bases to get the exact color I'm envisioning. Application can take up to an hour, as I experiment with various patterns and hues. Sometimes I need to scrub down with cold cream to bare flesh and begin all over. One shaky hand with a dark lip liner pencil and I have to start from scratch. The vodka helps here. By now I'm on my second glass of ice. When I'm satisfied with my face, I press together three pairs of dark fake eyelashes and gently glue them into place.

Jewelry next. As further proof that God may in fact hate homosexuals, he's cursed me with the absence of any real earlobes to speak of. I have a vast clip-on earring collection, and nothing to actually clip them on to. I get around this with generous dabs of spirit gum, which adheres them like cement to my earlobes, but removal becomes a battle frequently lost by several layers of my flesh.

My plastic breasts have been sitting on the counter since the beginning of the process, filled with water slowly coming to room temperature. I used to use a net to gather my fish from the aquarium resting between the two sinks, but my fish have long grown accustomed to my hand and now swim into my cupped palm to be transferred into the breast. When one needs to be replaced, I generally give the amateur a few weeks to grow accustomed to the rest of the troupe before calling on him for his debut appearance.

I struggle into whatever outfit I've chosen and slip the breasts, and mirrors, and tiny flashlights into the holes on my chest. My outfits are all my own designs, sketched with mechanical precision and then passed on to a girl I met in Atlanta when I first started doing drag. She's a genius tailor. She knows all my measurements and can work with any material I dream up, from stretch vinyl to faux leopard fur. It's better than Christmas when I come home and there's a package from her waiting for me.

A matching decorative corset comes last. I've mastered the elaborate contortions required to alternately loosen and tighten the laces behind my back until I've squeezed six inches of flesh off my waist and into various nether regions of my ass, thighs, and chest. Sitting is not an option for several hours. When I hail a cab to whatever venue I'm expected at, I have to enter sideways and lean across almost the entire length of the back seat.

The wig is always the finale. Until then I still feel like a boy playing with his mother's makeup. Granted, she'd have to be a pretty gaudy mother, like a single PTA mom trying to steal someone's husband.

I wrap toilet paper around my fingertip and wipe off the foundation in a clean line along the edge of my wig cap. Then I apply a thin streak of spirit gum along the hairline and quickly settle the thick, heavy wig into place before it dries, pressing the edge firmly into the glue. Into the back I press bobbie pins through the webbed cap of the wig, through the nylon wig cap, and try to lodge a few into my short-cropped natural hair. Anyone can wear a wig out for an evening, but keeping one in place through hours of dancing and twirling is a true aerodynamic engineering feat.

A quick stage check of whatever flashing lights or mechanical aspects my particular outfit requires is next, followed by the packing of my bag. Emergency touch-up makeup, a small roll of duct tape, cab money, gum, and I.D.

By now I'm usually up to an hour late, but there's always time for a quick swig from the bottle in the freezer before heading out. I pace for a bit, maybe dance a little, trying to work out any stiffness in the costume, or predict any potential dislodgement disasters. Fluid movement is the final piece of the illusion. Many men have experimented with dressing up in women's clothing, but no matter how successful with their look, a swaggering gait will shatter their efforts.

After hundreds of nights of lengthy and painful preparation, I've reached one solid conclusion. I would never, ever, ever want to be a woman. Aqua, on the other hand, has been sneaking up on me since childhood, demanding to be set loose.

The Tunnel used to be a garage for subway trains on the west side, but now its cavernous, labyrinthian rooms serve as a final destination for thousands of clubgoers each weekend. No one save the cleaning crew must really have a true grasp of its immense layout. With multiple entrances and exits, dozens of huge rooms open onto other large spaces, which in turn open onto yet others. Upstairs, downstairs, basement chambers, it's almost impossible to even gauge how many stories are in the building.

Being inside is similar to being inside the mind of a drunk or drug-addled partyer. It's a mirror facing a mirror, the space reflecting the flowing, amorphous, infinity-pushing thoughts of its patrons. One room thundering with dance music inhales and exhales its inhabitants into the next room, with a different DJ and décor and vibe. The Kenny Scharf Lounge, named after its designer, is covered floor to ceiling to walls with fake fur and glowing orbs. Sometimes it's used as one of the VIP rooms, although there are three or four more that can be pressed into service if needed. The main dance floor remains industrial, looking much as it probably did when lines of train cars were parked there, except with racks of stage lighting and audio equipment lining the ceiling and walls. Off this room are four other small spaces with bars and DJs, and the same goes for upstairs and downstairs as well.

The frontal lobe of the operation is up a small set of stairs just inside the Twenty-seventh Street entrance. At the top of the stairs is an unmarked door that opens onto a suite of offices that look like any offices one would find in the wholesale district. Plain gray metal desks and filing cabinets line the walls with slightly hipper than average office workers sitting around adding things, handing out pay envelopes, and filing papers. This is where I head at the end of the night to pick up my couple hundred dollars cash.

But that's a long way off. Four and a half hours, to be exact. Four and a half hours, twelve vodkas, three makeup reapplications, and two quarter-sized blisters on my feet away from home. Some nights fly by like Christmas morning, and others drag on like a family reunion. Tonight is the latter.

Jack, Ryan, and Grey have come with me to work tonight. It had been a long week at the ad agency, having just won a piece of the ABC News account, so Jack and I hadn't seen much of each other all week. Hence this rare trip out together. He doesn't go out much at night anymore, except for his calls, having had his fill of it when he moved to New York six years earlier. While attending grad school, he hung around with a group of performance artists who held their shows at dingy little East Village clubs. One of his favorite performances involved him lying down nude on a table while audience members were invited up to lick and attach postage stamps to him anywhere they wished. One of his artist friends was the person who initially steered him toward the escort business as a way of making extra cash.

Much of my busy work week had been spent at the ABC *World News Tonight* studios and offices. It had been Laura's and my ingenious insight that ratings might be improved by giving the viewing public a glimpse at the exciting behind-the-scenes workings of Peter Jennings et al. So the two of us tried to be as inconspicuous as possible lurking around the area the journalists called "The Rim," filming the reporters and editors with a handheld video camera.

Peter sits at a round table with his editors and previews stories as they come in from reporters around the globe, occasionally rewriting sections himself, typing away with two fingers at the computer stationed behind him. What was initially pretty fascinating quickly grew mundane, and we soon found ourselves trying to get shots of Peter Jennings picking his nose or topless as he changed shirts before going on air. I can safely say if we were trying to remedy viewers' perception of a dry and boring newscast, they weren't going to be that much more titillated by the lack of shenanigans going on behind the scenes.

It was a slow news week, and I started making deals with God that I would stop drinking so much if only He would conjure up a small tactical nuclear war somewhere in the world. I spent an inordinate amount of time reworking my fantasy about Anderson Cooper showing up in the ABC cafeteria and asking me if I wanted to come up to his office for a quick drink and blow job. I had gone in expecting wacky newsroom hijinks, since my entire experience of TV news comes from an unhealthy obsession with Mary Tyler Moore reruns. But instead I was stuck, day in and day out, with a bunch of behind-the-scenes old white guys trying to mug for our cameras.

The days' only excitement for me came at six fifteen when Peter disappeared into his office to get made up for the broadcast. I would grow more and more nervous for him as the live six thirty start time approached, wanting to scream, "Come on, Pete! Get your groove on! My mom and dad are wrapping up dessert, flipping channels to ABC *right now! Let's hustle, buddy!*"

But Peter always got into the studio just as millions of aging people were clearing away supper dishes and tuning in to see what news ol' Pete had decided they needed to hear about tonight. Laura and I sat in the control room and watched the broadcast, joking with the technicians during the commercial breaks filled with ads for various liniments and nutritional supplements. By the end of the week, the news seemed anything but.

In the fluorescent-lit basement room that serves as a dressing area at Tunnel, I'm mixing a fresh batch of glittering gold lip gloss to replace the layers I had left behind on my last four vodka glasses.

"Here, hold this," I say to Jack, handing him the gold pigment while I measure out the correct amount of gloss with the concentration of a biochemist.

"One of your eyelashes looks loose," he says.

"Thanks. You haven't had any calls yet?"

"I had one, but they didn't want to pay enough. I'm having a good time here."

I'm not sure if I believe him, but I'm glad he's here with me anyway. It's a big night at the club. Grace Jones is scheduled to give a small charity performance at three, and the club is packed. Personally, I'm not sure which is more charitable . . .

someone giving Grace Jones a gig, or the audience having to stick around to witness it.

The dressing room is crowded with other drag queens and go-go boys. Some work for different promoters, and others, like me, work for the club itself. Getting a regular club gig is the best a drag queen can hope for. Very rarely do we have to actually perform; mostly we just get to show up week after week and act like we're having the time of our lives. We're like the professional laughers that TV shows hire for studio audiences.

"Are you ready to go back upstairs?" Jack asks. "Ryan and Grey are dancing."

"I've just got to change my batteries and I'll be up. I'll meet you in the main room," I say.

"What if I get a call?"

"Just tell any drag queen—they'll let me know."

Jack leaves and I start the intricate process of retrieving the tiny flashlights positioned in little pockets all over my costume that make it glow. I'm just drunk enough that my fingers have abandoned all pretense of dexterity.

"Need a hand?" It's Carlos, a cute Puerto Rican go-go boy who has a little crush on me. He's just finished smearing glittered body oil on himself, and he has that alluring sexy smirk that all guys whose brains aren't wired to handle any sort of deep complexities in life have. Carlos is mostly straight, and during the day he works on his uncle's construction crew in Long Island, but at night he works the gay clubs because they pay more than straight ones, and he gets much better tips. Plus, he likes going home with an occasional drag queen. Or two.

"Sure. I can't get the one in back, right there," I say, pointing to the light attached inside the waistband of my miniskirt. Carlos kneels down behind me and reaches up my skirt. All airs of modesty are disposed of in the dressing room. We're all wizards here, and we're all behind the curtain. Nothing to hide in Oz. He maneuvers in the fresh battery I hand him and stands up in front of me.

"One hand scratches your back, you wash mine," he mangles in his heavy, incredibly sexy accent. He's pointing at his G-stringed crotch. He's in need of refluffing.

While the guys who get picked to work as go-go dancers unfailingly have some sort of genetic predisposition toward genital elephantiasis, they all still augment their gifts with the act of fluffing before they go onstage. This involves the manual stimulation, either by themselves or a nearby volunteer, of their cocks into a semi-turgid state. Once the desired girth is reached—and no extra since New York State law prohibits the display of full-on hard-ons—a rubber band is slipped around the base of their penis and behind their balls like a cock ring. The extra blood that has flowed into their dick is thus trapped there, ensuring an impressive bulge in their G-string—and more numerous tips.

"Okay, Carlos," I sigh, as if being asked to take out the trash. "But only because you're so cute."

I sit down in front of him and with one hand I pull down

the front of his G-string and with the other I grasp his cock and softly start stroking it.

"No," Carlos says. "With your mouth."

I'm about to point out to him the intricate perfection of the lipstick I just reapplied when he grabs my hair and pulls my face toward his crotch.

It's Sophie's choice. My wig or my lipstick. Since I have no extra hairspray in my bag, I dive headfirst onto his dick with the resignation of a child being forced to eat his vegetables before dessert. Not that it's a wholly unpleasant affair. Plenty of people in this club would say I have one of the best views in Manhattan at the moment.

While working on the task at mouth, I glance from side to side wondering if any of the dozen or so other people in the room are taking any notice. Of course they're not, and I find myself a little disappointed. Am I not being sexy enough? I return to the job with renewed determination. I even let out a moan or two, as if I were in an open audition for a low-budget porn movie.

My redoubled efforts seem to pay off. The drag queens around me start to cheer me on, and I smile as widely as I can considering the situation, and give a beauty queen style wave to my newfound audience. With one free hand I grab my makeup brush and pretend to reapply my blush without breaking rhythm on Carlos. The room explodes with laughter, including Carlos, while I continue to touch up my entire face—eye shadow, powder, lip liner—all the while bobbing up and down on his dick. By the time I pick up my compact mirror and pretend to fluff up the bangs on my wig most of the room is doubled over in tears. The fish are sloshing around in a contented rhythm, everyone is having a great time, and Carlos is inflating at an alarming rate.

And then it happens. My right eyelash gets irretrievably enmeshed in Carlos's pubic hair. The lash had been coming off all night, so after Jack noticed it, I applied three times the amount of glue to it as I normally do. Now the combination of drying glue and tangled pubic hair has my face more or less permanently attached to Carlos's crotch. Something's gotta give, either my eyelid or a chunk of his pubes. I want to scream in laughter with everyone else, but having my mouth full, I'm reduced to a combination of chortling and gagging, which sets everybody off even more.

Ginny Tonic comes to my rescue with a bottle of spirit gum solvent, which winds up stinging my eyes so badly my tears streak my mascara down to my chin. But perhaps the saddest thing is that the entire ordeal was for naught. By the time my face is unimpaled from Carlos's cock, he's laughing so hard he's completely lost the stiffness I'd sacrificed so much for. I send him over to the corner to start over solo.

Back upstairs I start searching for Jack. For ordinary mortals, trying to find someone in a busy club is like trying to find a needle in a haystack at night during a lightning storm. The pulsing strobes illuminate the crowd just long enough to pick out a body part or two. Sound travels out of one's mouth mere inches before it gets knocked apart by a sonic bass beat. But to drag queens, this is our daylight. It's like high noon on a clear day.

I see Jack by the champagne bar and sneak up behind him.

I reach around and stick my hands in his front pockets. He spins around and tries to kiss me.

"I wouldn't," I shout, "you have no idea where this mouth has been."

"Do you want some water?" he asks.

"Only if it came out of a fermented potato," I reply.

"You're not still drinking."

"I have to. I wouldn't want all these drink tickets to grow up with self-esteem issues."

"Just pace yourself. You've still got three hours to go," he says.

He's right, of course. But this is no place for temperance. Ryan and Grey come up beside us, both sweaty from dancing.

"What do you guys want?" I ask them, holding up three drink tickets and, before they can answer, add "and get me a vodka while you're at it."

Some people have mantras; I have a coda. I can find a way to end most sentences logically with "and get me a vodka while you're at it." If my last name wasn't already hyphenated, I'd consider legally adding that phrase to it. "Kilmer-Purcell-And-Get-Me-Another-One-While-You're-At-It."

The night has turned around successfully for me. By the time I'm halfway through with this seventh drink, I'm reaching the zone. Time starts slipping, and I have a hard time remembering the evening as one long narrative. It's breaking up into little moments with little or no connective tissue. One moment I'm with Jack, Grey, and Ryan at the bar, and the next I'm giving a lapdance to a guy in a wheelchair who I think has cerebral palsy. When one moment grows dull, there's a better one in the room next door. Standing in line in the unisex bathroom I spot Andy Dick. Minor celebrities always welcome the company of drag queens. It gives them the added sparkle that they haven't quite earned themselves. Suddenly the thought occurs to me that there would be nothing funnier than to be able to say that I touched Andy Dick's dick, so I take him into a stall and do just that.

I flit from room to room, corner to corner, dancing on top of speakers and throwing candy from balcony railings. Each segment of the evening is a succinct little capsule of uninhibited fun, fueled by vodka and the fact that there is no actual progression of time anymore. Word has spread through the club that Grace Jones won't show up for a while yet, which surprises no one and only exacerbates the tide of gossip about her massive drug problems.

When I reach this point, even the conscientious eleven-year-old Episcopal altar boy in me shuts the fuck up and starts to have a good time.

• • •

Eva Corvetta and I are doing a mock lesbian sex show on a speaker when Tony, a go-go boy wearing only a towel, comes by to tell me that Jack got a call and had to leave. The go-go boy towel dance usually comes about two-thirds of the way into the evening. They each mount a speaker and dance while dangling a skimpy white gym towel in front of their dicks. Male dancers can't legally show their genitals, but this segment at least gives the audience the illusion of the possibility of getting a peek. Eva and I pull Tony up onto our speaker to dance with us. We take turns lifting the corner of his towel and peering underneath, acting out theatrical reactions of amazement for the crowd. Being a drag queen in a loud club requires much the same dramatic skills as those of a silent movie actress.

Soon we're sandwiching him between the two of us, rubbing and grinding, of course always mindful of protecting our costumes and makeup. With him safely hidden between the two of us, we pull his towel off completely and take turns spinning it over our heads.

Okay, it's safe to say that nobody dons a pair of metallic opera gloves to open a stubborn jar of pickles. They're slippery. And this is why, completely unintentionally, while I was spinning Tony's towel over my head, it *accidentally* slipped through my fingers and flew out into the crowd.

There was very little chance that anyone in the crowd was going to find it in his or her heart to return it.

This is how Eva and I wind up ten feet above the dance

floor with a hot naked man sandwiched between the two of us and about four hundred drunk, high, and horny gay men between him and the nearest dressing room. I could go to a thousand wars and never see the look of fear in any man's eyes like there was in Tony's. The dance floor surges toward our speaker, cheering in anticipation. Eva and I do the only thing two compassionate drag queens could do at a moment like this.

We jump down off the speaker and leave Tony standing by himself.

With drag queens, the audience is always right.

12

I double-check the address that Jack had written on the scrap of paper and head up the brownstone stairs, being careful not to catch my heels on the long white fake fur cape dragging behind me.

I left the club early to make it to this party. Jack told me to try to be there by three a.m , and it's now three twenty-four, according to my cell phone.

Apartment 4A. I hit the buzzer.

"Griffin," the intercom crackles.

"Chinese delivery," I shout into the box as I'd been instructed. The door buzzes, and I go in.

It's trashier than I'd imagined. The hallway is carpeted in a threadbare mauve rug, and the walls are painted the same glossy putty color as most mid-priced rental buildings across the city. The stairs are covered in brown linoleum tile, chipped and peeling in the corners. The staircase tilts away from the wall as they do in all buildings over fifty or so years old.

Listening to Jack call into this place from our apartment led me to believe the headquarters of his escort agency must be at least as impressive as our gleaming penthouse. But now confronted with the dim reality, I realize that operating an illegal high cash flow business probably calls for a certain level of deceptive discreetness.

When I reach the fourth floor, I hear the muffled laughing and low music from behind the door at the far end of the corridor. I reach for the knob, and I'm surprised to find it turns easily and opens into a bright fluorescent-lit space decorated in a fashion not too different from a doctor's waiting room. I would have thought I needed a secret knock or further password or something.

Instead, the dozen or so people in the room, mostly women, turn to look at me, a moment of confusion passing their faces as they try to comprehend why a seven-foot-tall drag queen has joined their little party.

"Aqua!" Jack calls out from a sagging couch in the corner covered in a brown corduroy bedspread.

As soon as he calls out his recognition, the rest of the room relaxes and a few let out small chuckles of greetings.

The women all look similar, in an Eastern European

way—blond, thin, and leggy. Not quite pretty enough to be models, but more attractive than the average woman on the street. They're all dressed in a tacky style of evening wear, tight short black dresses made of cheap material and Payless heels. Most are smoking.

Jack rises to greet me, and Ryan and Grey wave from their spot next to the couch. In the corner, nestled between two tall gray filing cabinets, sits an older woman at a desk covered in magazines, Diet Pepsi cans, and ashtrays. The room reeks of too many varieties of cheap perfumes and sweat.

Jack hangs my cape on a hook by the door and ushers me over to the desk.

"Elaine, this is Aqua. My boyfriend," Jack says with his arm around the small of my back.

The folding card table next to the desk is covered with a paper tablecloth printed with pictures of balloons and confetti and the word "Congratulations" with three exclamation points. In the middle of the table, among plastic plates and forks and knives sits a half-eaten grocery store sheet cake. The section that remains uneaten reads "Hap . . . Retirem . . . Elai" in swirly red frosting.

"Hi, Sweetie," Elaine says in a cigarette-graveled voice, reaching out a hand. She looks to be about seventy, with hair the pallid shade of blond that only years of covering gray can achieve. "Aidan's told us all about you, but you're even prettier in person."

"Happy retirement, Elaine," I say, "what a great party."

"I keep waiting for my gold watch, but I guess I'll have to make do with this," Elaine says, holding up a gold Zippo lighter and matching cigarette case still nestled in its gift box. She's joking, but obviously tickled and proud of the gift. "Everybody chipped in," she adds.

A few of the other girls have come and circled around me, bending over to examine the fish in my tits.

"Are they real?" one blonde asks, tapping on the plastic.

"Are these?" I reply, smiling, grabbing her left breast.

"If you've got enough cash, they can be made out of anything you want," Elaine answers for the blonde. Everyone laughs.

I settle down on the couch next to Jack and two other women. Ryan, Grey, another male escort I haven't met, and another blond hooker with a bad nose job sits across from us. Elaine busies herself with trying to find a livelier station on the radio sitting on her desk.

"This is Tiffany, and Shelia," Jack says, gesturing to the girls on the couch, "and that's Roger and Tonya," he adds, pointing at the two others across the beat-up coffee table covered with old *Glamour* and *Elle* magazines.

"Hi," I say, "Hey, Ryan. Grey. What's up?"

"Slow night," Ryan says.

"Deadly," adds Roger.

Jack's told me that he doesn't like to hang out here, preferring to stay connected by beeper. Too desperate, he says, and now I see why. According to him, the only hookers who sit here waiting for calls are those who need cash from Elaine right away or those who have boyfriends or husbands who don't know what they do and would be suspicious of pagers. Besides, most of Jack's clients are his own private customers; he only gets two or three calls a month from the agency. The agency lets him call in to use their credit card machine when he has a customer who can't pay in cash. For a cut, of course.

"How long have you worked here?" I call over to Elaine.

"You know how they say this is the world's oldest profession?" she says. "Well, who do you think sent Eve to Adam?" She barks out a gruff chuckle at her obviously often-rehearsed joke.

Jack brings me a slice of cake, too sweet and bland. Dry. The edge that was exposed is especially stiff and stale.

"Can I get some booze?" I whisper to Jack.

"There isn't any. She doesn't like the girls to drink," he whispers back. "Just follow Tonya when she heads to the bathroom; she's got some in her bag."

The group of us around the sofa exchange small talk, Elaine chiming in with her jokes in between answering the phones and paging other hookers. Occasionally Elaine motions for one of the girls in the room to join her at her desk. She scribbles an address on a sheet of paper and goes through the terms that she'd just settled over the phone before sending the girl on her way. Often the girls make a stop in the restroom to apply a little more makeup on top of their already tricked-out faces before they depart.

"How long do drag queens have to work before they retire?" Ryan asks me kiddingly.

"We don't retire, we spontaneously combust on top of a speaker one night, showering the crowd with clouds of glitter," I reply.

What will all of us be doing fifteen years from now? No one in the room, except Elaine, looks like they could possibly be older than thirty.

"I'm going to night school," Tonya says, "for landscape architecture."

I've learned that Roger and Tonya are a couple. Both are from small towns in Pennsylvania, and met here at the agency. Roger came in thinking he was going to take wealthy older matrons to dinners and benefits, only to be told laughingly that there is no such thing as the "gigolos" he'd seen on TV and the movies growing up. The very few calls that come in from women are nowhere frequent enough to make any sort of a living, and mostly the women just chicken out before the escort arrives anyway. Every male hooker, straight or gay, has to earn his money from men and the occasional couple.

Tonya knew exactly what she was getting into, being introduced to Elaine and the agency from an old roommate who'd since moved to Palm Springs to work with Heidi Fleiss and do a little porn. Tonya spoke of the roommate with a certain reverence, as if the girl had made it big. Tonya had an air of resignation about her, as did the rest of the girls. They did not live a life of limos and champagne as pictured in the Yellow Page ads the agency ran.

I look around the brightly lit room. I hadn't had that much to drink tonight, and the fluorescent lights sober me up more than a hundred coffees and cold showers could.

"How long do we have to stay?" I whisper to Jack, just as a heavyset bearded man walks through the door. He looks about fifty and stops just inside to light a cigarette off one of the escort's lit cigarettes. He is wearing a cheap acrylic golf shirt and ill-fitting brown slacks. He mutters greetings in Russian or Polish to a couple of the escorts as he makes his way over to Elaine. He ignores Jack, Ryan, Grey, and Roger, only acknowledging the women. He stops when he gets to me and looks me up and down.

"What's this?" he asks in a thick accent, to no one in particular.

"Hey. I'm Aqua," I reply.

"You working?"

"Just visiting," I say. "For Elaine's party."

He doesn't answer, and instead picks up some papers sitting on the desk and rifles through them.

"Have you ever thought about working?" Tonya asks. "Stand up a second."

I do, half hoping that Jack will follow suit and we could leave.

"Turn around," Roger says. "You've got a great ass. I'm sure you could pull in some cash from trannie chasers."

I sit down again.

"You'd have to lose the fish," Tonya adds. "Guys like it real. I know a girl who works at a shemale agency. You could give her a call."

Before I can answer she's scribbling a number down on a corner of a magazine page.

"This is Rog's and my home number; give us a call tomorrow and we'll hook you up."

She hands me the ripped scrap, and Jack takes it and stashes it in his pocket.

"We should all get together and have brunch sometime," Jack says, changing the subject.

"Ladies and tramps," Elaine yells out, standing up as she hangs up the phone with an exaggerated flourish. "I've dealt with my last prick. Literally."

Everyone laughs and she stands and takes a pink cardigan sweater off the back of the desk chair and folds it into a small ball. She slips it into a brown paper shopping bag with handles that's packed full with personal effects from the desk. The stern man replaces her in the desk chair, and all the escorts stand up to form a receiving line of hugs and kisses as she makes her way out of the agency.

"And who says whores won't kiss?" Elaine jokes, reaching the door. Jack and I follow her out the door, and Jack takes the heavy bag from her as we head single file, with her in the lead, down the stairs. In the dimly lit stairway she seems as if she's shrunk five inches as she stiffly and awkwardly takes one step at a time down the four flights. I watch the back of her head as she hobbles down, trying to see through her skull to the thoughts she must be thinking as she leaves four decades of organizing paid sexual encounters behind her for good.

Out on the street she turns to walk east with us down Fifty-fifth Street. She takes Jack's arm. Suddenly she's the same as any of the hundreds of old ladies walking around New York's sidewalks on warm autumn nights when the weight of accumulated years makes it impossible to sleep. I fall behind the two of them.

"Thanks for stopping by tonight," she tells Jack.

"It was a great party, Elaine," he says. "I'll miss talking with you."

"Oh, I'll page you every once in a while in between *Wheel of Fortune* and *Jeopardy*. Gonna have to get used to falling asleep when everybody else does."

"You going to take a vacation or anything?" Jack asks.

"I don't know, maybe visit my sister in Palm Springs. I might move there one day," Elaine answers.

"I grew up in Southern California," Jack says.

"I can tell. You're an open person. Like talking to the desert." She turns to me. "Watch this one," she says, nodding toward Jack. "He seems simple, but you just never know."

I laugh, not knowing how to respond. Luckily she spots a free cab heading up Third Avenue and her attention turns to reaching the curb before it speeds by.

Jack hands her the bag.

"Keep in touch," he says, helping her into the back seat.

"Happy spanking!" She smiles back as she closes the door and leans forward to give the cabbie directions.

The water nearly overflows the tub as Jack climbs in behind me. I lean back onto his chest and he wrings out the washcloth of warm water over my head. I close my eyes as the water runs down my face, relaxing away the forced smiles and clownish faces I'd had to pose all evening. My cheeks sting from the nightly scrubbing away of thick foundation. He rewets the washcloth and drapes it over my chest.

I take a sip from the vodka he's set on the side of the tub for me.

"Let's retire," I say.

"Okay."

"Where should we go?" I ask.

"We'll find a little hacienda on the ocean in Baja."

"I want to grow lemons," I say.

"And we'll get a goat."

"And we'll open a little bar in a nearby village where Aqua can sing standards and make people cry for lovers that they can never have," I say.

"She'll ride a burro back and forth from home every night, and all the village children will follow her with flowers."

"And she'll toss them Chiclets and Lifesavers."

"I'll make you profiteroles and serve them to you on the beach every night," Jack says.

Jack pours shampoo in my hair and starts rubbing my head. It smells like lavender.

"I'll build us a raft. We'll lie on it watching stars and memorizing constellations," I say. "And I'll learn Spanish and sing you to sleep."

He starts singing softly to me and rinses my hair with clear water squeezed out of the washcloth. I'm dozing off when he lifts me up and wraps me in a thick white towel and guides me to bed.

I just begin to fall asleep when I feel Jack, curled behind me under the cool sheets, tracing out letters with his fingertip on my bare back.

"I love you more than heaven," he writes softly, letter by letter.

13

Jack and I watch *Blue's Clues* every morning. Or at least every morning that we're both at home.

We both have a little crush on Steve, the host. We lie in bed, me usually still a little drunk from the night before, and Jack winding down from his multiple nighttime calls. Something about Steve makes everything about the world seem manageable. His simple soothing sentences make sense to me during that short wistful window between inebriation and hangover. For Jack, they lull him further and further away from a surreal night of deviant sex and violence. Even when Jack has an overnight client in the apartment, we find time to retreat to the bedroom and lie on the bed silently watching Steve, speaking only when it's time to solve the day's puzzle.

"What time's your flight?" Jack asks me when the closing theme song strikes up.

I roll over on my stomach, drape my arm across his chest, and mutter into my pillow.

"Five o'clock."

"Do you have to go into the office today?"

"Just for a little while. I need to show a rough edit to a client," I say. "Will you be home around two before I go?"

"Probably, unless I get a call."

Still on my stomach, I pivot around until my legs are over the side of the bed and slide limply off onto the floor, where I lay flat on my back looking at the ceiling.

"I don't want to go," I say. "I'm too tired. Don't make me go."

Jack leans his head over the edge of the bed and looks down at me.

"Shut up, dickwad," Jack says. "You know you want to go. It's a free vacation."

"Vacation? You try hauling five wigs and six outfits halfway around the world. And how the fuck am I going to buy goldfish in Tokyo?"

"They're called *koi*," Jack adds helpfully.

"Great, I'll wander around the city in black Spandex and heels, waving cash and pointing at my tits while yelling out '*koi? koi?*'"

With his head directly above mine, Jack proceeds to pretend that he's going to let a gob of his spit drop down on my forehead.

"That drips on me and your teeth will be the next thing falling out of your mouth," I say.

Just as it looks like it's going to fall, he sucks it back up into his mouth.

"Pussyboy," Jack taunts, smiling.

"You five-dollah-whore, I ten-dollah–dlag queen," I say with a bad Hollywood Japanese accent.

"Get your flat ass off the floor and start packing, round eyes," Jack says, getting up, prodding me softly with his bare foot.

"Pletty geisha boy take hot shower now," I say, continuing with the lame accent. "You want soap me? Ten dollah you touch bottom place."

"Two dollah, final offer," Jack says, pulling me up toward the bathroom.

"Ow. You likee smackee smackee? Five dollah you spank me."

I get into the shower and Jack brushes his teeth. When he's done, he turns around and puckers his mouth up against the glass shower door. I kiss him from the other side.

Last month I performed at Wigstock, the world's largest, and arguably most important, outdoor drag queen festival. Tabloid television crews from all over the world descended on the west side pier where it was held and filmed segments for their shows back home. One Japanese crew was particularly fascinated by me and my costume. I posed for endless shots and stilted interviews. I later realized their interest lay mainly in my tits, the goldfish being a symbol of good fortune to them.

Shortly after, I got a call to entertain a group of Japanese businessmen at a karaoke bar in a basement in midtown. It was a birthday party for one of the executives at the Japanese television network I'd appeared on. Though of course I remember little of the evening, I apparently was a huge success. The next day I got a call asking if I wanted to go on a short ten-day tour of Tokyo.

Everyone at the ad agency was tremendously obliging, even though I'd already used up all my vacation and sick days. Once again, my appeal as an agency mascot trumped my value as an actual contributing employee. One of the partners even gave me a substantial pocketful of spare yen that he'd collected over years of Japanese business trips.

My itinerary is a bit unclear to me, having been relayed in a phone call that came in the middle of one night last week. From what I can tell, it involves a couple of clubs, a television appearance, and, I think, a wedding reception.

By the time I return from the office before heading to the airport, I'm more nervous than I thought I would be. This is my typical pattern. Rush headlong into adventure and then dig in my heels right before going off the cliff. I would give anything to be one of those people who just do without thinking. Like Jack. I have too many years of being the good boy behind me not to be aware of potential pitfalls ahead. Thankfully I have my good friend, vodka, to help me, sometimes literally, stumble off the edge of the cliff. I'm slowly donating my liver to the pursuit of finding my balls.

Jack's not in the apartment when I get home from the agency. On the dining room table he's left a note with an intricate illustration of my plane trip drawn in the margins. A tiny jet swoops past the Chrysler Building, Big Ben, and the Eiffel Tower, before skipping across the bottom of the page over the Pyramids and the Taj Mahal and winding up back at the top circling Mt. Fuji. Little goldfish jump through the clouds, and the blank spaces in between the famous landmarks are dotted with cactuses. In tiny little capital letters he writes:

HEY GEISHA GUY, GOT A PARTY CALL. PROBABLY A BIG ONE—DAYS. I'LL BE ON YOUR PLANE, AT YOUR HOTEL, AND IN YOUR DREAMS. SEE ME IN EVERYTHING, AND HEAR ME ON YOUR SHOULDER. TELL AQUA TO COVER HER TITS BEFORE SHE EATS SUSHI.

GO IN THE BEDROOM CLOSET IN THE FILING CABINET, BOTTOM DRAWER. UNDER THE MAGAZINES, THERE'S A METAL BOX. TAKE WHATEVER CASH YOU NEED. I MISS YOU TO PIECES, JACK.

Jack knows I'm in a perpetual state of poverty. Having to do three or four shows a week means constantly updating and creating new costumes. I've been told I'll be getting a pretty big sum for this trip, but I don't know if I'll get it when I arrive or after I've finished. I'm nowhere near as adept at negotiating payments as Jack is, and sometimes I'm so drunk at the end of my shows that I forget to pick up my cash altogether.

I give Jack a good portion of my agency check as a token rent payment. He didn't want to take any, but I told him I didn't want to feel totally kept. *Ho-dependency,* Laura calls it. He spends far more on me than the little I give him, and we both pretend not to notice.

The gray-green metal box is exactly where he indicated, and it is much heavier than I expected. When I was little I had a box just like it where I kept my rock collection. Random pieces of quartz and mica that I'd picked up on the side of the gravel road that we lived on. Shiny things. Glittery things. I'd wash them regularly under the garden hose to make them glisten even more. Make them more precious.

Jack's box was even heavier than my rocks. I lift the lid.

"Holy fucking Jesus damn Christ!" I think to myself. Maybe I even say it out loud. Who could tell? If a tree falls in the middle of the woods next to a huge goddamn pile of cash does anybody hear the fucking tree? Inside, packet after packet of hundred dollar bills bundled in their bank wraps are stacked neatly on their sides, filling the entire box. I counted how many bills were in one wrapped bundle and multiplied it by the number of packets. Normally I can't even add the hours on my time sheet correctly, but when staring down at this wad of cash I suddenly turn into the Rain Man of tallying up.

There's $357,000 and change. I didn't know this much cash existed in the world. Well, I knew, but I just never thought I'd see it in one place. What the hell does he have all this money for?

Then I put it together. Jack can't have a bank account. He doesn't really even exist on paper other than his birth certificate and college diplomas. He'd told me that he hadn't filed taxes in years, 'cause he has no legal income to report. If he gets the flu, he goes to the emergency room, gives a fake name, and leaves right after they give him a prescription. He makes cash payments to the previous owner of the condo, a guy who also was involved in the escort business and had to leave town and start over in Palm Springs. Jack doesn't even have a checking account and has to get money orders every month to pay Con Edison.

It's too much cash for me to comprehend. How many asses were beaten with leather straps to make all this? I find myself unable to take a single hundred dollar bill. It's simply overwhelming. The idea of money for sex now has a physical size and shape and smell and it lives in our closet. I don't feel any real repugnancy to it, just a newer greater awareness. Like finally seeing a bruise on the skin after feeling a little bit sore for a day. I close the lid. I put it back under the stack of old issues of *GQ* and close the drawer. I shut the closet door and get my bags and lock the front door and hit the elevator button and hail a cab and go to Japan.

I wish I'd never seen it, and I'm not sure why.

14

Here's what I don't like about Japan: there are thousands of social rules that you don't know about but are made to feel like you're breaking continuously. And at the same time, everyone is too polite to tell you.

I'm a freak here. In or out of drag. I'm being put up in a private members-only hotel where only one person claims to speak English, but notes are continuously slipped under my door telling me to be ready in an hour. No matter how many times I tell people that it takes at least three hours to get made up, one hour after receiving such a note a teenage boy name Toshi shows up at my door smiling and telling me, "Now we go."

Whoever brought me here, to their credit, had a check

waiting for me in my room upon arrival. Unfortunately, not having a bank account in Japan has left me unable to cash it. Thus, I'm trapped in my room ordering fishy breakfasts and watching Japanese TV until the times arrive that "Now we go."

The first three attempts at sending someone to buy goldfish for my breasts resulted twice in koi the size of my forearm, and once with some sort of fillet fresh from the store. After many crudely drawn pictures and charade attempts, I've now got four healthy small goldfish swimming around in the sink in my bathroom. The idea of successfully communicating my need for a fish tank is beyond my wildest jet-lagged dreams. I resign myself to brushing my teeth in the tub.

After finally figuring out the phone, I try to call Jack. No answer. I leave a message. One of those pathetic messages that go on and on with a half dozen aborted wrap-ups because I can't bear to break the tenuous connection with home.

"Now we go," Toshi says at the door.

"You go hell," I reply, smiling. Toshi smiles back and waves me into the hallway.

Vodka's not so easy to come by here. I manage to physically intimidate or possibly just confuse Toshi enough to procure me five bottles from the bar at the wedding reception I emcee. And by "emcee" I mean talk onstage to myself while two hundred Japanese people have no idea what I'm saying. Luckily, the drag queen crotch grab is indeed a universal joke.

The television appearance seems to go better, but really, how would I know? From what I can gather, it's some sort of talk variety show with an androgynous host who shrieks something that sounds like *"icky bicky koon hiiiiiii,"* which makes everyone in the studio audience respond with a hearty *"Hi Hooooo!"* and dissolve into raucous laughter. After what I think is an enthusiastic introduction, someone pushes me out onstage, which I assume means I am to start my number.

I'm wearing my leopard print cat suit, with studded collar and wrist bands. I have a cat-o'-nine-tails, which I occasionally threaten the front row of the studio audience with. While I'm lip-synching to a song called "Twiggy Twiggy" by the Japanese group Pizzicato Five, a video montage of New York City rolls on the screen behind me. The scenes of the Empire State Building and the Brooklyn Bridge are intercut with random scenes of Japanese people eating ice cream cones. I don't know why. I suspect neither does anyone else, but, hey, *icky bicky koon hiiiiiii!*

The rest of the trip is just nightclub after nightclub. At one small club, referred to by Toshi as a hostess bar, I notice a man behind the bar arguing with my trusty chaperone. The room is only about the size of my old studio apartment and is located on the thirtieth floor of a swanky high-rise. The showy light displays on the signs of the Ginza district shine below us. From this height it looks exactly like Times Square. It's strictly a club for drag queens and their admirers, and I'm joined by a dozen or so other Japanese drag queens and transsexuals whose attention to gender-bending detail eclipses mine by several degrees. They're so small-boned and frail I can't begin to even imagine them as men. They're simply beautiful, moving around the room with the small graceful movements that are the hardest part about mastering gender transformation.

Toshi comes out from behind the bar and approaches me, smiling as usual. He points to the drink in my hand.

"Sure, I'd love another," I say, holding up my glass.

"No. Too many bar buy," he says, smiling.

Great. Getting cut off in any language sounds dismally similar.

"No. Man buy you," Toshi continues.

Laura had joked about me getting sucked into the international sex slave trade before I left, but given how I had won over the hearts and soul of Japan on television days earlier, I thought I was in the clear. I'm about to scream out *"icky bicky koon hiiiiiii!"* to create a diversion and run when an older Japanese businessman comes over and sits next to me.

"My name is Mr. Hatsumoto. I am happy to see you here," he says politely, bowing his head.

"Aqua," I say, bowing back. At least my potential new master is polite.

"What is trying to be said," Mr. Hatsumoto continues, "is that in this bar, you are to compel gentlemen to purchase drinks. They are one hundred dollars each."

Christ. For a hundred dollars back in New York he could get a drink and a hand job. With plenty of change.

"You gotta have quite a thirst to buy yourself a hundred dollar drink," I say as coyly as I can.

"Or a lot of hundred dollars," Mr. Hatsumoto winks back.

Three hundred dollars and half a buzz later, Mr. Hatsumoto turns out to be a pretty swell guy. Not my type at all, but he doesn't seem to be hitting on me anyway. I find out he supports three of the girls in the room, paying their rents and giving them spending money. The girls swing by occasionally to check on him like he was their grandfather, and he buys them a drink and requests songs from them.

"Do you have a gentleman?" he asks me later in the night.

"Yes, I do," I say.

"Does he watch you?"

"Yes. Very well."

"He is a fortunate gentleman," he says.

"That's what I tell him," I say.

"I could not live without my girls."

"Sounds like they wouldn't do very well without you either," I reply.

"We survive together. We give pieces to each other. I give more. I have more. But they have more valuable," Mr. Hatsumoto says.

"I'll take your word for it, Grasshopper," I say, not sure if it's the booze or the conversation that's causing me to drift.

"Now, I will take you back to your hotel," Mr. Hatsumoto says, rising from his seat.

I'm having a fantastic time at the club. It's homey. All the girls and men know each other, have been together as a group for years. From what little I understand of the girls trying to speak English to me, most of the men here are married but have their "girlfriends" at the club. Some wives know, some don't. Some of the men are horny pigs, and others just fond admirers. Like I've learned from my nights in New York, and my life with Jack, once you've crawled into what's commonly thought of as the sordid underbelly of life, you realize it's all just different versions of normal.

I wonder what Jack's doing. It's nearly four in the afternoon in New York. I haven't reached him on the phone once since getting here. He left me one message at the hotel early in my stay, "All well. Miss you lots. Aqua too." But I haven't heard from him since.

More than anything I wish he were here with me. "A relationship is an accumulation of shared history," he'd said to me once. And here I was making history without him. It's lonely. And I can't wait to go home. Parts of me are showing through my Aqua, and I'm having a hard time keeping them separate.

"You have given me what you have to give. Thank you. Please let me take you back," Mr. Hatsumoto says again.

"Okeydokey. Now we go, Cricket," I say to Toshi, collecting our things and following Mr. Hatsumoto out the door.

I have seen what I needed to see, and am ready to go home. To Jack.

15

The apartment is a shithole.

Every window is wide open and the frigid October wind is howling through, pushing everything that isn't heavy enough into swirling piles in the corners. A Celia Cruz CD, Jack's favorite, is blaring on the stereo.

"Jack?" I yell out. No answer.

I put my bags down in the foyer and walk into the living room. His backpack is lying on the floor, flayed open with bottles of lube and dildos and leather toys spilling out like guts.

"Jack? I'm home."

No one's in our bedroom. The faucet is on in the master bath and I walk in and turn it off. There are scorch marks on the countertop and bottles of lotion and Vaseline scattered on the floor. The floors in all the rooms are covered with muddy footprints. The candle on our nightstand has burned down completely and spilled a puddle of wax onto the unmade bed.

Back in the living room there's a stack of foil deli takeout containers in the corner filled with barely eaten egg burritos. The futon is folded out flat and dragged into the middle of the room. It's covered in stains. Sheets have been dragged out of the bedroom and lie wadded up in piles next to it. Every towel we own is scattered around the floor.

On the plane on the way home, I didn't have a drink. Not one. A twelve-hour flight. It was the longest I'd gone without alcohol in as long as I can remember. I am tired of always being drunk or being sick. Of not eating. I am tired of throwing up when I do eat. And I missed Jack. Entirely. I thought of his note. *See me in everything, and hear me on your shoulder.* The entire time I was in Japan I *did* see him everywhere. "Does he watch you?" Mr. Hatsumoto asked me. "Yes. Very well," I said back. And it was true. When Mr. Hatsumoto told me to leave the club, I did. Because Jack was watching. Mr. Hatsumoto was standing in for Jack. *Pack it in, Aqua. Time to go.* Aqua fading out, me showing through. A class-A, diamond-studded fuck-up. I can't tell when it starts to happen; someone needs to show me. Make me notice. Pull the plug when everyone else wants me to keep drowning. When everyone else wants me to show them how to have a good time. Wants to push me under and dance on my head. The geeky first chair junior high bassoon player peers out from behind the false eyelashes, pleading for someone, anyone, to pull the fucking plug. Pull the goddamn fucking plug. This is Jack's job.

Where the hell is Jack?

A message on the machine:

"Aidan, it's Dolores." Dolores is the new dispatch operator at the escort agency in midtown who replaced Elaine. "Where are you? I have a party in a room at the Regis. They need party favors and two guys. Open ended. Call when you get this; your pager's not working."

Where the hell is Jack?

Where the hell is Jack?

Where the hell is Jack?

I don't know what to do.

Where the hell is Jack?

I stand motionless in the foyer staring at the door. I've never felt more sober and more drunk at the same time. But I'm not drunk. I'm just confused. I haven't talked to Jack in ten days, my home is a disaster zone, and I'm confused.

I turn around in circles, trying to spot something that makes sense.

Jack's pager is on the kitchen counter. Seeing the pager not attached to Jack is like seeing someone's lung sitting on the counter with them nowhere in sight.

As I step into the kitchen, shards of glass crackle under my shoes. Broken crack pipes. I pick up the phone.

Hey, it's Ryan and Grey, we're not home, leave a message at the beep.

I don't know what to do.

I don't know what to do.

Where the hell is Jack?

I don't know what to do.

Please. Please. Come in the door. Come in the door and pull the plug.

I go to the freezer and pull out the bottle of Absolut. This is all I know how to do when I don't know what to do next. The bottle freezes to my hand as I walk back through the windy living room.

I take a long swallow from the bottle and set it on the floor. My stomach grinds in opposition.

Little by little, through the night, I pick up pieces of my home and put them back where they belong.

By the time *Blue's Clues* comes on, the apartment is back to how I left it.

Please come in the door.

Please come in the door.

BOOK

III

16

Why didn't you call me? I told you to come over," Laura says over my shoulder as I finish an ad layout on my computer.

"I thought for sure he'd be back. He's never been gone this long without calling."

I've been home from Japan for two days. Still no Jack. His pager goes off incessantly through the night. I can't drink enough or pass out hard enough not to hear it. Yesterday at the office I called it myself a half dozen times, hoping maybe he'd come home and reclaimed it.

"Do you want me to sleep at your place tonight?" Laura asks.

"You're a cheeky filly, aren't you? You know I don't like women."

"Not half as much as they don't like you," Laura says. "Is the money still there?"

"I didn't check."

"Do you have the numbers of any of his regulars?"

"No."

Laura stands behind me, watching me work over my shoulder. I've worked on the same layout for a day and a half. Try the logo in this corner. In that one. Maybe a little bigger. Try orange. Part of my indecision is just nerves, and the other part is because I've kept a steady buzz going since the night I got home.

"Have you been drinking scotch?" Laura asks, wrinkling up her nose.

"I ran out of vodka."

"You stink."

Even I can smell it on my skin. I've avoided any meetings unless they're in a large enough room that I can sit far away from anyone.

"You should at least take some of his money and get yourself some more vodka," Laura says.

"You should write self-improvement books for the successful drunk," I reply.

I'm supposed to work at Barracuda tonight. The prospect churns my stomach. Nearly as much as the prospect of going home and waiting in the apartment for another night. I try not to think about it. I'll just keep shuffling the logo around this page like it's an oracle that, once I find the right combination, will give up an answer about what I should do next.

Laura and I have been given the best assignment in the agency. An antidrug campaign for ABC Television. If we can sell an idea it'll mean at least a month in LA producing spots with celebrities from the network. Unfortunately we have only a week left to come up with something. I was supposed to be thinking about it while I was in Japan.

"Maybe I'll just come to your place and we can work tonight," Laura offers.

"I can't, I've got to do Aqua."

"Great. Another wasted day tomorrow. If we get fired, I'm stealing Jack's cash," Laura says.

"Fifty-fifty. Deal."

I can hear the salsa music playing as I get off the elevator in my building. Joy and fear well up inside me in equal portions.

When I reach the door, I hear the vacuum running inside. I step inside and see Jack at the far end of the apartment vacuuming the living room. He's naked and his back is turned toward me.

I stand just inside the door and watch him. He's singing along with Celia Cruz and pushing the vacuum wand back and forth along the parquet. I remember the day I moved in. After all my boxes were safely in the apartment, he started playing this CD as loud as it would go. The salsa mixed with the heavy summer air coming in through the windows, and he grabbed me and we danced around the piles of boxes and furniture.

"Hey!" Jack shouts, noticing me in the reflection of the bay window. He turns around with a wide grin on his face. "You're home!"

I have no idea what to do. Absolutely none. I don't even know what I'm feeling. I freeze. To buy a little time I start taking off my jean jacket. Still nothing comes to me. I try to dredge up the excitement I felt two days ago in the cab on the way home from the airport. Nothing. I try to remember the waves of anger I felt over the last two days. Nothing. By the time I'm reaching for a hanger, I'm completely empty.

Jack's turned the vacuum off and is walking toward me.

"When'd you get home? I thought you came home tomorrow," he says.

"I got home two days ago."

Jack throws both his arms around me and lifts me up off the floor.

"Where were you?" I ask.

"Huge party. Huge. Eight straight days. Twenty thousand."

"Your beeper was here."

"I know, we had to move everyone here for a while after the guy got kicked out of his hotel. He was still paying, so we didn't want to end it."

It's a simple explanation. The same one I'd come up with in my head several times over the last two days. Exactly what I'd expected. Big party, then Jack, Ryan, Grey, and the other escorts go hide out a few days to recover. Why am I such a paranoid freak? I make fun of people like me.

"How was geisha boy? Are you a superstar?" Jack asks.

"Pletty goldfish rady big hit," I say. "They were already talking about me coming back."

"Not too soon," Jack says.

"Don't worry. Me have prenty full of dirty rittle Japanese businessmen." I head into the bedroom.

"Where are you going?"

"Aqua's on tonight. Barracuda."

"Shit. I was going to take you to dinner," Jack says.

"No dice. I've been gone from New York the equivalent of four years in drag-queen time. Gotta work on my comeback tour. Bring me a vodka and we can talk while I get ready."

That first night I was home from Japan I'd put the apartment back together with Germanic precision. Everything tucked into the drawer or closet it belonged in. But I hadn't unpacked Aqua. Her three big bags were shoved under the bed, repacked in the same mess that I left them in after I had to explain them at customs coming in.

Jack watches me pull her out. I restyle the wigs and smooth out the costume I need for tonight. Slowly I feel myself coming back. It hurts a little. Like when I would come in from sledding in the below-zero Wisconsin winters. Mom would make us sit in a lukewarm tub as we slowly came back to room temperature so our feet and hands wouldn't burn unbearably as the blood returned to them.

When I'm in the shower shaving my legs, Jack comes in and straddles on the side of the tub, watching me. I tell him about the TV show, the wedding, eating salty fish for breakfast in my modular hotel room. I tell him about the hostess bar, cute little Toshi, and the kept girls. I start to tell him about Mr. Hatsumoto, but stop. I can't remember enough.

By the time I'm heading out the door we've totally reimagined out lives with me as Japanese celebrity sensation and Jack opening an escort service to cater to the multitude of Japanese business tycoons who no doubt crave a little humiliation.

"Leave me a note if you get a call, okay?" I say, kissing Jack and closing the door.

17

It's just the TV," I tell my mother on the phone. "Hang on a sec."

I cover the mouthpiece and throw up again in the toilet. I find the idea of trying to throw up as quietly as possible kind of funny.

"Why don't you turn it down?" she asks.

"Jack's got some people over; they're watching a movie."

Every word that comes out of my cotton-filled mouth requires Herculean effort.

"How was LA? Who did you meet?" she asked. My mother's voice has always sounded like an eight-year-old girl's on the phone.

"Drew Carey, Bob Saget, a bunch of people. I don't know," I say.

"Anybody on my soap?"

"I told you, it's a campaign for ABC. Your soap is on CBS."

"Did I do something wrong? Are you mad at me?"

I consider letting loose with the litany I've collected of things she's done wrong over the years, but hold back. I wouldn't be able to make it halfway through without vomiting, and I want to save it for a time when I'm a little more oratorically suave. Besides, they're mostly pretty petty transgressions, anyway.

"No, I'm not mad. We're just in the middle of the movie. It's at a good part," I say.

"Well, I can tell you don't want to talk, so go back to your movie. Say hi to Jack."

"I will."

After I hang up I decide to lie on the bathroom tile just a little longer. I'm relatively certain I'm not expected anyplace, and right now the less I move the less I hurl. It's about the most complex equation I can deduce about my current situation.

I have no idea what's going on in the living room, except that it's loud and there's more than one person. Why must mornings always be such a puzzle?

It must be Saturday, since Mom always calls at the beginning of the weekend. I know I got back from LA on the redeye last night. We filmed celebrities making little fifteen-second antidrug speeches on a sound stage in Ventura the entire previous week. It was like school picture day. One by one they would arrive, get made up, sit in front of a backdrop, and blather on about how drugs are bad, they fuck up your schoolwork, mess up your family, blah, blah, blah. One of them even showed up sky-high. We couldn't use anything we shot of him.

Laura and I spent our nights partying at the Skybar. I brought Aqua with me just for fun, and the second-to-last night we were there we invited the client out with me in drag. She was young, and we thought she could handle a little fun.

When the news reached the agency back in New York, they freaked. Apparently, there's a fine line between being creatively interesting and potentially client-threatening. News to me. Two of the partners wanted me to come home immediately. Which I would have, had they been able to find me.

Laura, the client, and I stayed out the entire night and went straight to the shoot the next morning, me still in drag. My plan was to go to the wardrobe truck and find suitable clothes for the day, but when the director saw me he wanted me to shoot a spoof spot for the campaign. When any drag queen is faced with a camera, we become powerless to stop ourselves. Any resulting catastrophe can hardly be blamed on me.

The difficult part of improvising an antidrug spot when

one is still completely smashed is coming up with one's motivation. After a few insincere attempts I came up with this:

> Look at me. I'm beautiful and you know it.
> Using a whole bunch of drugs didn't get me to where I
> am. Using the right bunch of drugs did. Kids, be responsible.
> Know your junk and talk to your dealer.
> This message brought to you by:
> ABC TV. Trying Really Hard to Care.

Now it's not like they're going to air it or anything, so I'm really not sure what the resulting fuss is about. By the afternoon when we all sobered up, the client didn't think it was that funny anymore and begged us not to tell anyone back in New York. Unfortunately, Laura and I had been on our cell phones all morning telling everyone we knew.

By the time we got on the plane home that night the client had turned a little chilly toward us. Luckily, before we checked out of the hotel, Laura and I had collected our entire minibars to keep us company on the flight home.

I got home from LaGuardia at six forty-five in the morning and walked into our apartment in the middle of another party. They've been happening with increasing frequency during the past month. One client in particular wants to party two or three nights a week. Tired of getting kicked out of hotels, the group now meets at our place. Our kitchen gets turned into a crack lab, with freshly cooked rocks drying on a plate on the counter. Jack makes them smoke in there too, as a courtesy to me. I have never tried crack, and it doesn't really interest me. Even though his clients and escort friends are all well-heeled, I still have an image of crack as being the drug of choice for the South Bronx set. Anyway, I'm still perfectly content with my booze and very occasional sniff of old-school powdered coke. And maybe a little weed. Just an old-fashioned guy, I guess.

I use the handles on the cabinet under the sink to pull myself upright. Slowly standing, I give myself a quick once-over in the mirror. At least I got most of the way out of Aqua last night before passing out. I grab the phone and head out into the apartment.

Three guys are fucking on the floor near the living room windows. I watch them. They're oblivious to me standing twenty feet away. I'm jealous. Jack and I have only had sex once since I got back from Japan. They shift positions stiffly, mechanically; they've probably been fucking all night. I've learned that about crack from Jack. He says it makes people want to fuck for hours, mostly just to expend the nervous energy.

I walk into the kitchen to hang up the phone. Jack is there with an escort named Trey whom I met at my birthday party. I'm not that fond of Trey, for no particular reason. He just reminds me of that kid in high school who was always called in to the principal's office for questioning no matter what the crime. He could start a second profession as an extra in police lineups.

"Hey," I say.

"Hey," Jack says. I can't tell if he's high.

"Where's the client?"

"He's gone. He left last night. How was the show?"

"I don't remember. I'll wait for the reviews," I answer. "Can I turn down the stereo?"

"Sure. Go ahead."

Back in the living room I turn off the music completely. The fucking trio doesn't even pause. This is the part of the day when I'm most uncomfortable. I'm too sick to go out and do anything, and don't really have anything I want to do anyway. I don't have a show tonight. I have nothing to get ready for. I don't really feel like watching TV. I have no hobby. Except for drinking.

I start to wonder what the people all around me do on days like today. I live in a box in the sky with boxes identical to mine on all sides, and forty stories of identical boxes below me.

What do all those boxed people do on Saturdays? Do they go to movies? Museums? Do they walk their dogs? I see people with dogs on the street all the time. Why don't we have a dog? I used to go to brunch. Are people going to brunch right now? Or does everyone wake up as trashed as me and find their spouse in their luxury kitchen with whores and crack? I want to peel up the floorboards and peer down at them. Get some ideas.

The thought of standing here all day simply waiting for the next throwing-up episode starts to crush me.

I throw on a pair of sweatpants and my down jacket and head for the door.

"Where you going?" Jack yells from the kitchen.

"Out to walk the dog," I say.

I walk the streets the entire afternoon, trying to unravel the mystery of Saturdays. I head down Madison Avenue, staring at couples, trying to see what's in their shopping bags. I see people eating in windows, getting into cabs. It's all I can do to stop myself from running up to them and asking, "What are you doing today? Where are you going? Can I join you?"

I try Laura on her cell. Maybe she wants something to eat. I'm sure she's home, but after this crazed past week in LA she probably sees my number on her caller ID and she doesn't pick up. I stop in a diner near Twenty-fourth Street and order a tuna melt but have to leave halfway through because I can't keep any of it down.

I keep walking. Down Madison to Twenty-third. People coming out of a movie. I follow two women down Park Avenue and try to listen to their conversation. One of the women is debating with the other whether or not she should start trying to get pregnant. At Fourteenth Street I run into the farmers' market. Hundreds of people are picking their way from stall to stall, inspecting apples and squash and onions and other things that remind me of the garden I had growing up.

I pick up an acorn squash and decide to buy it. It's reassuringly solid and heavy as it swings back and forth in its plastic bag, knocking against my leg as I walk on. Maple syrup. I used to love squash baked in the oven with maple syrup. I pick up a bottle and add it to my bag. Look at me. I'm shopping. Butter. I need some butter too.

It's almost six o'clock by the time I get home. I dread going inside. I half-expect the same robotic threesome to still be fucking in our living room. But they're not. The apartment's empty. It's a wreck again. I can't keep up, and Jack and I keep forgetting to call the cleaning woman.

I'm too tired to bother with the squash, so I just take it out of the bag and set it on the counter. That's enough for me, actually. Just looking at it there on the counter amid the piles of tin foil and burnt spoons and Brillo pads and glass pipes and baking soda and rubbing alcohol and the rest of the crack paraphernalia is soothing enough.

Jack's asleep in the bed. Without taking off my clothes I crawl into bed next to him. He's hot, and sweaty, and oily. He's frowning in his sleep. Jack crashes hard after a party, and then beats himself up for missing workouts or time with me and his friends.

He wakes up fitfully as I crawl up closer to him. His skin tastes like crushed aspirin all the time now.

"Hey, you," he says groggily.

I pick up the remote and turn on the TV.

"Hi."

"Where you been?" he asks.

"Grocery shopping," I say, staring at the TV.

"Admirable."

We watch *Sanford and Son* for a while. Silently.

"Jack?"

"What."

"I think we should slow down our partying for a week or so."

"I know," he says.

"I'm tired."

"You're drinking too much," he says. He's right. I'm now way past the point of it being a silly little vice.

Sanford and Son ends and *Andy Griffith* comes on. Jack gives a halfhearted attempt to whistle along with the theme song.

"Are you having problems with crack?" I ask him during the first commercial break.

"Maybe a little," he says.

"I thought you faked smoking."

"I usually do. There was just that one big party while you were in Japan. It went on too long."

"You suck when you're high," I tell him.

"Sorry."

"Let's take a week off. No rock. No calls. No booze. No Aqua," I say.

"Have you asked her?" Jack asks.

"She'll use the time to go to Mexico and get that abortion she always wanted," I say.

"Okay. A detox week."

When *Andy Griffith* goes off, I flip to a nature show. Snakes of the Amazon.

"Everything's cool?" Jack asks me.

I think for a sec. No. Everything's not that cool, really. But what other option do I have but to try for a week to make it cool again?

"Yeah," I say finally.

"Good."

I roll on my side and close my eyes.

"We're having squash tomorrow," I say before drifting off to sleep.

18

It's a week of cactuses and hearts.

Jack sends me to work every day with a note or little drawing in my pants pocket. The partners at the agency decide they aren't mad at me anymore when one of the suits at ABC decides my spoof was hilarious. Someone at ABC who knows someone at CBS says there's even a possibility Letterman might run it as a parody of the campaign.

Jack spends his days working out and buying things for the apartment, and then meets me after work to have dinner with Ryan and Grey or Laura. Thursday he spends the entire day shopping at Mexican markets and putting together a three-course Mexican dinner for me. I know he's just trying to stay busy to keep his mind off of getting high.

I'm scraping leftover bits of enchilada off a plate into the trash when he comes into the kitchen and puts his arm around me.

"You know, if we wanted we could move to Baja Peninsula tomorrow and get a place on the ocean," he says.

"Not enough work for drag queens. Or whores," I say.

I instantly regret using that word. It used to be a funny running joke, but this week it sounds dirty. Mean. Jack had gone the entire week without taking a single call simply because I asked him to. When we first started dating, he grilled me to make sure I had no problem with his career, and I assured him I thought it was cool.

"I'm sorry. Stupid dig," I say.

"Don't worry."

"I was just joking."

"Don't worry. I know. I've decided, when I get back to work next week, not to go on any party calls anymore. Just normal ones," he says.

"You don't have to do that. You faked it before, you can do it again," I say, trying to be overly generous.

"Maybe later. Not now."

I'm relieved. I know that he couldn't be around crack any more than I could be around an open bottle without indulging.

I'm also not naive. I'm not sure he can do this. Sex for

money, dressing in drag, and too much booze . . . fine. All things we can handle. But the crack throws me. The whole time we've been together I knew that we were forging a different path for ourselves, pretty far outside the typical romantic comedy genre. But genres are genres for a reason, and I saw enough "very special" sitcom episodes about the dangers of drugs to know there aren't a lot of happy endings. No one made it through puberty in the 1980s without Nancy Reagan's harpy message permanently tattooed on his brain. Every time I saw Jack high I couldn't help but picture his brain as a sizzling egg in a frying pan.

But if anyone can break the cliché, I suppose it might as well be us. Whatever fucked-up lifestyle we've been living, it's had its positive effects. I've been getting fewer and fewer doubtful snide remarks about our relationship from people around me. They knew what I was like before Jack and they see what I've been like since. I might still be having a bit too much fun for some of them, but they know that since Jack, I've been showing up to work when I need to and have been nowhere near as bitchy. Something about Jack is good for me. Something's working. Maybe not all of it. But something.

"Something wrong? You look a little sober," Laura says Friday morning. She's been teasing me relentlessly all week.

"Don't you get a pin or something after your first week sober?" she goes on.

"I think I get a drink ticket," I say.

"When can you start drinking again? You're beginning to bore me."

"You've been boring me forever. Why do you think I drink?" I reply.

We're sitting in her office avoiding any sort of productive thinking that might possibly lead up to an advertising concept for Kudos granola bars, our new assignment. So far, we've come up with one brilliant tagline that was summarily dismissed by the rest of the agency:

Healthy granola. Sinfully delicious chocolate. Kudos . . . it's bi-snack-ual.

"I want you to come out Saturday night and meet this guy I've been seeing," Laura says.

"As much as I want to witness a miracle, I can't drink till next week."

"Bring Jack. He'll stop you."

"We're going upstate for the weekend, anyway."

One of my surprises from Jack this week was a reservation for a weekend getaway to a spa two hours north of the city. He decided we both needed a little professional help in finishing up our little self-imposed detox. A couple of seaweed wraps, salt scrubs, hot stones. Total cleansing.

"If you make it through the weekend without drinking, I'll take you out to get trashed at lunch on Monday," Laura says.

"Deal," I say. "And if you can make it through the weekend without ditching this new guy, I promise not to tell him about your herpes."

"I don't have herpes."

"He doesn't know that."

"Don't cross me, motherfucker," Laura says, turning back to her computer. "I'll crush you."

Again, the salsa music when I step off the elevator. For weeks after coming home from Japan I braced myself for whatever possibly could be behind my apartment door. But tonight the only thing crossing my mind is whether to pack my Gucci or J. Crew bathing suit for our spa trip tomorrow.

"Hey!" I shout over the music.

A cupboard slams in the kitchen. I walk around the corner. Jack and Trey lean against opposite counters staring at me bug-eyed. Jack's head is completely shaved.

"Hey, you!" Jack says. "How's work?"

They're high. The kitchen smells like someone peed on a pile of aspirin then lit it on fire.

"Fucker," I spit.

"What's wrong?" Jack starts, before he realizes he's not going to get away with any lie. "We just had some left over. No big deal."

I don't say anything. Jack's chapped lips had just begun to heal this week and now they're blistered and swollen from the burning pipe. I notice a burn that goes from his right temple over the top of his ear. He must have flared a chunk of his hair off, then shaved the rest of it off. The angry red splotch has Vaseline smeared on it.

"Nice haircut, asshole."

"Come on. I'm sorry. This is it. It was the end of it. I wanted to get it out of the house."

I stand there and glare at him. He looks like a seven-year-old. He starts to open his mouth and say something, then stops. Then does it again. Like a fish.

I've never seen Jack this soon after he's taken a hit. He's wild with energy. Even though he's standing perfectly still, his body is in constant motion. His forearms twitch. His hamstrings are contracting. There's no part of his body that's not completely taut.

"You told me last week that there wasn't any more in the house," I say.

"I forgot I had some in my backpack."

"That's bullshit. You cleaned out your backpack on Tuesday." Jack spent most of the evening on Tuesday taking inventory and rearranging his work toys.

I nod at Trey. "*He* brought it over."

Trey looks at me and smirks, shrugging his shoulders. Then he turns toward the counter and picks up the pipe and the lighter we use to light the grill on the balcony. Trey lights up again, and I see Jack's attention waver between the pipe and me. And it looks like I'm losing.

"Fucking weak-ass lying crack whore," I spit at Jack and storm into the bedroom.

By Sunday night I've only been home a total of three hours. Just long enough to shave, change into a different Aqua outfit, glare at Jack as he lies in bed watching TV, and head out again.

It's good to be back to normal. My own peculiar normal. Working, drinking, working. I'm in great form at the clubs, and everyone seems especially glad to see me. Unlike most of my binges, I don't even bother to count how many drinks I'm having.

It's not easy to stay out twenty-four seven. You have to choreograph your club schedule carefully to always be in one that's not only open, but crowded and fun as well. It's best to find a pack and move together. The hardest time is between noon and about five p.m. Then I'm reduced to heading to my favorite illegal afterhours club on Avenue D. It used to be called BodyHeat, but after a string of overdoses it's now referred to as BodyBag.

BodyBag is located in an apartment on the first floor of a tenement building. It's an old railroad-style layout. To get in, Baron, the guy who sits on the stoop, has to have seen you before. But even if you're there every night, there's no guarantee that he's in a mood to let you in. It's nothing personal. Baron keeps a close watch on everything that happens on Avenue D.

If too many people are coming and going from the place, it looks suspicious and he'll just wave you on when you approach. There's no point in stopping and pleading; it will only guarantee that you won't get in the next time.

Apparently the coast is clear when I arrive sometime Sunday afternoon from wherever I was last. I try to remember where I just came from. I can't. Doesn't matter. The only thing that matters is that I'm away from Jack and on to the next party.

Inside, there's a smattering of tired partyers scattered on the frayed couches and armchairs. The windows are all painted black. I have to pee, but there's no use trying the bathroom. It's always filled with either people preparing to shoot up, or people shooting up, or people passed out and blocking the door after shooting up.

The kitchen area of the apartment has been turned into a bar. I walk up and pay twenty dollars for a double bottom-shelf vodka.

As soon as I sink into one of the chairs that reeks of the sweat of the last hundred clubgoers that collapsed in it, I start to pass out.

Keep it together. Just finish this drink. Then I can go home and pass out without having to talk to Jack. One more drink. That's all.

I lean back and shut my eyes and listen to other conversations around me. I can't follow any one for any length of time. People's voices come in at different volumes and at different intervals.

"... I don't have any. Bob did it all before we left ..."

"... fucked me till I bled ..."

"... right out the goddamned window ..."

"... she's a fucking whore ..."

"... beat the crap out of her ..."

"... nice fish ..."

"... I said, *nice fish.*"

Someone's poking my shoulder. I open my eyes. A huge muscular bald guy is sitting on the arm of my chair.

"Nice fish." He's pointing at my tits. "Want another drink?" Now he's pointing at my glass.

"Sure. Vodka."

"Be right back."

I'm spinning again when he returns.

"Guzzle and let's go fuck," he says.

I close my eyes and swallow my drink.

I can't stand up.

"Here." He's holding out his hand. He pulls me up and I lean into him. We pause for a second. I swallow hard. I try to even myself out somehow. I try to take a deep breath. Fucking corset. He puts his arm around my waist. It's thick and weighty. I barely have to move my feet as he lifts and pulls me toward the door.

"It's not free," I say when we're out in the hallway, "I'm not free." My eyes won't open the whole way.

"What do you want?"

"Three hundred dollars."

"I'll give you forty dollars and a rock," he says, waving for a cab.

"Hola, senorita."

Pedro. I don't answer him. I just need to get this over with. I try to keep my focus on the elevators at the far end of the lobby.

"Nice place," the bald guy says, gripping my arm tightly. Too tightly. But it's keeping me upright.

In the elevator he pushes me against the wall and shoves his hands down the back of my skirt. They're huge hands, and rough. I bury my face in his chest. His shirt smells acrid, like our kitchen. One of his hands comes out of my skirt and pushes into my forehead, slamming my head into the side of the elevator. His tongue is in my mouth. It's sour. I can't breathe.

Inside the apartment I put my bag on the kitchen counter. Just want to rest a second. Jack's cleaned it off. No crack shit anywhere. I lean my forehead against the cupboard next to the stove.

"I need the money first," I say.

"Come on, baby. It's no big deal. We're here now, let's just go."

"I need the money."

"Look at this place, bitch, you don't need my fucking

money," he says, laughing, though I can tell he's losing patience.

"Just the rock then. Just gimme the rock," I say.

"Here," he says, pulling a vial out of his pocket. "You want to hit it now?"

It's hard to think. I could do it. See what it's all about. Maybe if I tried it I would know what Jack was up against. Nah. I'm fucked up enough. I just want to get this over with.

"Let's just fuck," I say. I tumble forward and fumble for his belt. He spins me around and shoves me. Hard. I fall into the hallway closet door and see a flash of black around the edges of my eyes. For a second I think I can recover and stay upright, but the wall's not where I judged it to be. I land on the ground face first.

He's on top of me pressing his huge forearm down on my cheek, smashing my head down hard against the parquet.

I throw up.

The vomit in the back of my throat and nose makes it impossible to breathe. I gag again. His arm presses harder into the side of my face while he pulls at my clothes.

His knees are on the back of my knees, crushing my kneecaps underneath them. He's saying something, but my ears are ringing so loud from the pressure of his arm that I can't hear him. I can't breathe. I can't breathe. There's a blue light flickering at the end of the hall. The bedroom TV. Jack is home. I'm gagging harder, trying to get air. The hard cold parquet feels like it's pushing up against my Adam's apple. He spits in my face. It's drips in my eyes and I can't see.

"Jack," I try to call out; it comes out as a hoarse whisper.

"Jack!" I say again. My voice is smothered, but a little clearer. I feel the vibration of my vocal cords against the floor. "Help!"

I gag again.

"Jack!" I spit some vomit out, and my voice is stronger.

The guy stops pulling at my costume. The pressure against my head lets up a little.

I lift my cheek off the floor.

"Jack! *COME HERE!*"

The bedroom door doesn't move. If he's crashed, he'll never hear me. The sliver of blue light keeps flickering. The guy is frozen on top of me.

"*My roommate's coming,*" I sputter to the bald guy, knowing that he wasn't.

"*What the fuck, bitch?! This is some fucked-up shit. Fuck off.*"

A sudden shove sends my head bouncing against the parquet, and the guy stands up.

I hear the apartment door open. I still feel the phantom pressure of the guy's weight on top of me, but I know he's gone. I feel the draft rushing over my legs from the outside hallway.

I throw up once more, jerking violently, before I pass out.

• • •

When I wake up, I throw up again immediately. I turn my head to see the clock on the microwave. 1:37 a.m. I turn my head back, and the blue light is still flickering down the hall-way. I close my eyes again.

The clock says 3:23. I've pissed on the floor.

The clock says 5:56.

The clock says 7:33, and a faint pink glow reflects off the wax on the floor. I'm woken up by children's laughter in the hall. When the bald guy left, he didn't shut the door. I can hear the rustling of backpacks and lunchbags as the kids tease each other on their way to the elevator. Though I'm facedown fac-ing the opposite direction, I can tell when they reach our door-way by the sudden silence. I am not ashamed. Lying there half-dressed and broken in my puddle of piss and vomit, I cannot dig up *any* emotion, let alone an appropriate one. The group of children shuffle off to the elevator, silent now.

I get up on my hands and knees and crawl into the kitchen. I stay on my hands and knees for probably fifteen minutes be-fore I can pull myself up. My throat burns. My tongue is so thick it fills the back of my throat. Just water. I need water.

I see the top of the crack vial sticking out from behind the coffee maker. It must have gotten knocked there. The bald guy left without it.

I have a hard time swallowing. After the first sip I lean over the sink and throw up again. Just bile. And threads of pink mucousy blood. My stomach feels like crumpled wax paper scraping against the inside of my abdomen. My knees won't bend.

I have to be at work.

I take a deep breath that smells like puke.

I pick up the crack vial and head toward the master bedroom.

Jack's still passed out on the bed. I don't know when he crashed. The television is still on, but muted. He's sprawled on his side on top of the covers with the television remote in his hand.

I take a pen and Post-it note out of the drawer on the nightstand.

"Here. I won something at the fair for you," I write. I stand the vial on the note and head into the bathroom.

19

I can't stand silence.

That's part of the reason I like clubs.

When I stand and dance on the room-size speakers, I can feel noise through the bottom of my feet. When the lights strobe to the beat of the music, I can see noise.

When I was a kid, my brother would go days without talking to me if I'd done something to make him mad. I'd say I was sorry over and over again for whatever it was. Then I'd get pissed and yell at him. Eventually I'd cry. And he wouldn't say a word to me until whatever arbitrary time came that he decided he would talk to me again. He wasn't being mean; it's just how he handled conflict. We're polar opposites. When he has a problem, he gives the world the silent treatment. When I have a problem, I give the world a sequined, star-spangled, show-stopping spectacular.

Jack hasn't spoken to me all week. The crack vial stayed on the nightstand with the note, exactly where I'd left it until I finally threw it down the trash chute a couple of days ago. It was a shitty thing to do to him, and I'm a little bit ashamed of myself.

I sleep in the guest bedroom. In the morning, he orders his own breakfast, and a half hour later I get up and order mine from the same deli.

I cancel my Aqua gigs because I've had enough drama for a while. And because I don't want to give Jack the satisfaction of watching my drunken messes.

When I come home from work, I hear him watching TV in the master bedroom. But I don't go in there anymore. From the guest bedroom I still can hear his beeper go off in the middle of the night. I hear him rustling around the apartment while he gathers what he needs, then the final zipping of his backpack before the click of the apartment door lock.

I'm not sleeping.

He's not away from the apartment long enough on any call to be getting high. Just the minimum hour, and I hear him come back in, thump his bag on the floor, and turn the TV back on in the bedroom.

I'm not drinking.

Laura's been asking me what I'm going to do. I don't

know. All I know how to do is drink and go out and come home and go to work. And I don't feel like doing it.

I haven't stayed home on a Friday night since I moved to New York, and tonight I'm lying in the guest bedroom watching TV. I've seen every late-night infomercial more than once, and even purchased some skin cream from Victoria Principal because I feel sorry about her career. I get up and stand at the window looking out downtown. I'm bored. I have everything and nothing to think about.

Back in bed I listen to every sound. The plastic tarp over the table on the balcony crunching in the cold wind. The two short clicks in the walls before the heat comes on with a low whoosh. I hear a constant bass hum all around—the nervous system of the building carrying electricity and gas and phone conversations to all our respective little boxes.

I listen to it all—the constant, the rhythmic, and the random.

It's hard to measure the night by sound, but it can be done. I know that when the traffic noise is quietest, it's about four thirty in the morning. I know that when the *Times* hits the door, it's around five.

Now the clock says it's morning: 5:45. But the November sky still says midnight. I hear the elevator ding twenty yards down the hall outside our door. Seven seconds later I hear his keys in our lock. Then his heavy backpack hitting the floor.

I hear the refrigerator door open, the unsealing vacuum wheezing as the cold inside air meets the dry heat in the apartment. A cupboard door. A glass. The crescendoing fizz of a new two-liter Diet Coke bottle opening.

It's a one-sided conversation with no one actually talking. I lie in the dark, close my eyes, and try not to listen to his movements around the apartment. These are the sounds of our life together before it got so messy. I want to say something back. Anything. Anything that sounds like things sounded last summer. Even just to myself. Just something out loud.

The inside of my eyelids turn pink. My door has been opened and the light from the hallway shines through them. I won't open them. There is no noise.

Like an eclipse, the world behind my closed eyes goes dark again for just one second before I feel a kiss on my right eyelid.

I keep them closed.

A kiss on the left one.

I open them.

Jack looks down at me. Then closes his eyes.

He leans over and puts his forehead on my chest and goes limp.

"*Blue's Clues* is on," he says softly into my T-shirt, his muffled voice vibrating only a half-inch away from my heart.

20

Contestant, make your choice, I think to myself stepping out of the elevator and walking down the hallway toward our apartment. Coming home these past few months is like a game show. I have no more information about what I'll find on the other side of my door than I do any other doorway in the hallway. Will Jack be high? Will he be alone? Will there be an orgy in the foyer? I may as well let myself into any random neighbor's apartment . . . it couldn't be any more unnerving than mine.

Contestant, do you choose what's behind door number 42A? Where the middle-aged couple lives, whom I met once on the elevator dressed for a formal evening out? Are they right now eating their dinner of broiled chicken breasts and instant rice pilaf without a single word passing between them?

Or do you choose 42B? Whose door is decorated with a dried wreath made up of magnolia leaves and baby's breath that seem incongruous with the intense dark-eyed veiled Arab woman I'd seen taking the trash out one day?

42D? The family with the brood of four blond children who witnessed me passed out in a puddle of my own sick?

I choose 42E. Our apartment. I'm getting used to going for broke. Betting it all in the bonus round.

I reach our apartment and prop my knee against the door while I dig in my satchel, looking for my keys. Just when I feel their jagged edges deep in the bottom, the door opens and Jack stands in front of me, naked except for a too-short navy blue tie that barely reaches his navel.

"Going for an interview?" I ask.

"How'd you like to make five hundred dollars?" he whispers.

Five hundred dollars is about the same as I take home all week at my job at the advertising agency, and far more than I'd been able to command as Aqua lately. With drag queens showing up in every magazine and on every daytime talk show, the level of subversiveness of the profession has decreased dramatically and the number of job entrants into the night world of gender-bending has multiplied exponentially. Anyone who can afford to buy a housedress on Fourteenth Street seems to show up at the clubs and declare themselves a drag queen.

"If it involves another man's dick, I can live without it," I reply, my one and only disastrous attempt at hooking still relatively at the top of my mind.

"Only indirectly," Jack continues, still whispering, pulling me into the foyer.

In the last couple of weeks we'd gotten ourselves as far back into normal as we could ever credibly classify ourselves. Jack's been true to his word of "no party calls," and Aqua's been on her best behavior. (Or worst behavior, if you're the kind who likes it when she blacks out on top of the club speakers.)

Peering past Jack, I see a video camera set on a tripod in the middle of the living room. A cable snakes across the floor and disappears into our bedroom.

"Do we have our own cable access show now?" I ask, whispering for no good reason other than the fact that Jack is.

"I have a client in the bedroom."

"Our bedroom?!" Jack and I agreed early on that he wouldn't do anything with his customers on our bed. It may seem like an insignificant bow to traditional monogamy, but to me it's as close as we come to a family value.

"He's not on the bed. He's in the chair watching me on the TV. It's a live feed."

"Oh," I whisper. In the last few months I'm easily placated by most any explanation. Rational or otherwise. If I'm told enough to survive the situation, that's probably plenty. Any more and I'll just be frightened. Or disgusted. "Whatever this guy wants, I'm not going in the bedroom with him," I warn Jack.

"You don't have to. He just wants to watch two guys. I was just paging Grey when I heard you at the door. I thought you might like a shot at the money."

"What do I have to do?"

"We just need to make out a little, then fuck."

"While he's watching," I clarify.

"From the bedroom," Jack adds reassuringly.

Jack's apologized for the incident with Trey in the kitchen, and we've since settled back into a congenial routine, one that reminds me of the first couple of months we were dating. After weeks of not talking, it seemed now, in some ways, like we had just met. But a different kind of first impression. More like we'd just met through friends who had warned each of us about the other one. Told us the dark secrets we should know about each other. And now we're not so much overlooking them as we are looking past them.

With this newfound knowledge of each other, we'd been able to return to our old comfortable habits, deli breakfasts, reading the *Times* to each other, chatting in the shower about our respective workdays. But we avoided sex. We were romantic to the point of being treacly, but actual physical contact seemed to overstep some newfound boundary in our relationship.

One of the few bonuses of Jack's profession was that it afforded us opportunities like the present one. We could ease back into a sex life via cash sanctioned role-playing.

"Why the tie?" I ask.

"You have to wear one too," Jack says. "Part of the fantasy."

"Do I get to meet the guy?" I always want to meet his clients. Partly out of fear that they're better versions of me, and partly out of simple voyeurism.

"At the end. He's going to interrupt us and yell at us," Jack says. "He's the principal. We're the prep school kids."

I'm disappointed. Not by the prospect of being watched while having sex and then being yelled at, but that I'm agreeing to participate in such a cliché fantasy.

"He won't cum on our chair, will he?"

"I gave him a towel. He said he's not jerking off anyway."

Standing in the foyer, Jack begins to undress me. A feeling of closeness comes over me like I hadn't felt with him in a long while. I remember a night when I was at camp the summer before sixth grade when the three other boys in my tent were planning on holding a girl named Jennifer underwater and taking the top of her bathing suit off. They waited until they thought I was asleep. I was too "good" to be included in their scheming. I rustled and turned over in my bunk to let them think they had woken me up, hoping they would ask me to join them, but each time they would stop whispering until they thought I'd fallen asleep again. I hated being the good boy no matter how much I was rewarded for it.

I'm not too good for Jack.

Jack strips me to my underwear and fumbles while knotting the blue boys' school tie around my neck.

"I'll do it," I say, taking it out of his hands. He smiles at me. I kiss him on the forehead. "What's my name, Aidan?" I ask.

"Jonah," he offers, pulling me into the living room.

He sits me down on the futon and steps out of frame. I try not to look directly into the camera. I wish I had a textbook or something. Wouldn't that be more convincing? Studying for an exam?

Verisimilitude doesn't seem to be a big priority for this client, I conclude, and start fiddling with my tie to have something to do with my hands. La, la, la . . . sitting in my underwear fiddling with my tie.

Then Jack walks back in and sits at the opposite end of the futon. Here we are. Just two schoolboys, sitting around in their ties and underwear. I'm wondering if I should whistle or something, when Jack slides closer to me. I pretend not to notice him, staring out the window at the darkening November sky. The Empire State Building is lit up with orange and yellow lights and I'm trying to figure out the significance of the colors when I feel Jack's hand on my thigh.

It's warm. And tentative. I don't remember the first time Jack actually did touch my skin. Was it that first night when I came home with him from the bar and he watched me discard Aqua and listened to me talk in the bath that he drew for me? Did he touch me in bed that night? The night I don't remember because I was so drunk I couldn't remember his voice on the phone the next day?

His hand moves to my cheek and turns my face toward his.

"Hey, Jonah."

"Hey, Aidan."

"I saw you in gym class today," Jack says, slightly louder than he needs to, being only inches away from my face. I realize we need to be heard by the microphone on the camera sitting ten feet away. It suddenly hits me. I'm in a porn movie.

"I saw you in the showers," I reply at equal volume.

"I hope no one catches us," Jack continues, rubbing his fingers across my right nipple.

"There's no one here but you and me," I say, improvising and picking up his cues.

"I've never done anything like this."

"Me neither," I add. My abdomen starts shaking as I stifle a laugh that I know would reduce us both to tears if I let it go. "The principal would *kill* us if he knew."

Jack plants his face on my shoulder, pretending to kiss it but actually trying to suffocate his own giggling. Pathos. I need to dredge up some pathos in order not to dissolve into a laughing fit. None of my roles in high school productions of *Bye Bye Birdy* and *Brigadoon* have prepared me with any particular methodology for this current performance.

"Would you like to suck my dick?" is the best I can come up with. Jack is on the verge of busting out laughing, so my lap is actually a good place to hide his face. I'm an improv *genius*.

Jack obligingly starts fishing underneath the waistband of my underwear. I lean back and watch him kissing my navel. I start to rub his hair. He looks up at me.

Instead of heading down, his kisses wend their way up my chest to my neck. I feel his soft hair rub against my earlobe. His breath reaches my ear.

"I've always loved you," he whispers, far too quiet for the camera to pick up.

In the months we've been together, I've consistently tried to figure out what makes him tick, with no luck. He shuns all my attempts at pop psychoanalysis by saying "I y'am what I y'am" in a bad Popeye impersonation. I've fully explored what attracts me to him: his ease, his directness, the fact that he does pretty much anything he wants without being crippled by a stack of built-up fears and anxieties. Like being at the eye doctor as they place different lenses over your eyes—"Is *this* better, or *this one*?"—being with Jack is as simple as finding the lens that makes everything suddenly come clear.

But what was I to him? I'd been so relieved that suddenly I could see clearly that I just ran around looking at everything using him as a lens. I'd been looking right through him. Whenever I try to imagine what it is in him that finds me attractive, all I can come up with is entertainment value.

Now, while fumbling around naked in front of a camera hooked up to a TV in another room in front of a man I've never seen, I wonder what we look like to this stranger. I wish I could somehow go sit next to the client and watch Jack and me naked together on the screen, to see how we come together as one unit, what it is that makes us a couple. Talk with this stranger about it. Quiz the guy. *What are you looking at? Describe this scene for me. What are those two people doing together? Why?*

But right now Jack's skin feels so good to me, I can't imagine detaching. We touch each other like we touch ourselves, so similar that sometimes while kissing his body I find myself inadvertently kissing my own arm without being able to tell the difference. He smells like honey and cinnamon, like the desert and cactuses he draws for me. I try to be the sky for him. Mist and rain and breezes.

As it grows darker, the orange and yellow lights from the Empire State Building color our skin and the room around us. The wind picks up in sudden strong gusts that only occur this high in the sky. Fat, heavy gobs of snowy rain hit against the window. Still we are together. Our bodies turn and roll and rub together as long as the invisible man pays us to. Our skin grows numb from the constant touch, the rubbing against the fabric beneath us. The tease of being on the edge of orgasm for so long is intense—a dull, bruising pressure building inside us. As if it would be impossible for us not to be touching. To pull us apart would be as sharp a pain as pulling off our very skin.

"What are you two boys doing!"

The door to our bedroom slams open. The noise startles us and we fly to separate edges of the futon as if we really had been caught. I'd completely forgotten that this man was in our bedroom. Watching. He looks to be around fifty, with dyed auburn hair combed with a perfectly straight side part. He's wearing navy blue polyester suit pants that are too tight and bunch up at the zipper and a light blue short-sleeved dress shirt with two large round sweat stains under his arms. His forehead shines with sweat and his skin blushes with rage. I'm genuinely embarrassed. Still incredibly hard, I cover myself.

"What are you two boys doing!" he repeats.

"Nothing," Jack answers, doing nothing to hide his stiffness. Now it's easy to tell who's the professional. I suddenly realize it's not the sex that's the difficult part of being an escort, it's the bits before and after.

"I said what are you boys doing?!"

I'm at a loss now. I'm in my recurring nightmare of being in a play without ever having seen a script. And either this man knows his part really well, or he's genuinely pissed at us.

"We were playing," Jack continues.

"What were you playing with?!"

"We were playing with our dicks," Jack says.

"And?"

"Our mouths."

"And?"

"Our asses."

"Stand up."

We both do as he says. I watch Jack. I'll follow his lead.

The raging man seems as if he could drop dead of a coronary at any second.

"Come here."

We slowly walk over to where he stands. Having been lying down for so long, I grow a little dizzy and grab Jack's arm.

"Don't touch him!"

I let go. Now I just want to hang on to him more out of fear than imbalance.

"Kneel down."

I shoot Jack a glance. I try to figure out how to say, "I'm not giving this guy head" with just my eyes. We kneel down. The man's hands lower and I grimace, hoping against hope that his pubic hair isn't the same fake orange shade as his head.

The man places one hand on the top of each of our heads.

"Our Father, who art in heaven . . ."

Jack joins in.

"Hallowed be thy name."

I follow suit.

"Thy Kingdom come, thy will be done, on earth as it is in heaven."

We repeat the prayer at least a dozen times before he instructs us to sit back on the futon. He disappears into the bedroom and emerges a minute later in his jacket and hat. His face now sports a broad friendly smile. A complete hundred-and-eighty-degree mood shift. He asks Jack for a pen and paper, and then sits next to us writing the address and phone number of the church in Brooklyn he belongs to.

"If you guys ever feel like it, give me a call and we can go together one Sunday. There's a great youth service on Thursday nights too!" he adds cheerfully before heading out the door.

Jack calls the agency and tells them to end the hourly charge. Five hundred dollars apiece. Just like that. Finally, I'm a successful hooker.

21

Thanksgiving.

I don't care what Butterball.com says, the hardest part about cooking the perfect Thanksgiving dinner is avoiding the splinters of broken crack pipes that collect in the crevices of the kitchen floor.

Even though Jack's been clean for three and a half weeks, I still wind up getting little shards of glass embedded in my feet every time I make a pot of coffee. I've cut back to two shows a week, and treat them like a real job. I show up on time, drink just enough to keep dancing, get my cash and go home. We're proud of how well we've gotten everything back on track. However surreal that track may be.

Last week, on our way home from a movie, Jack saw an ad for fresh turkeys in the window of the Food Emporium down the street. Without telling me what he was doing, he dragged me inside and placed an order for the largest one they could get.

"We're having Thanksgiving at our place," he said. "An old-fashioned Thanksgiving."

"With drag queens and hookers and cranberry sauce?" I asked breathlessly.

"Just like at Grandma's," he replied.

The Thanksgiving Day Parade on NBC is still playing on the TV set in the living room as I'm trying to find something large enough to cook the turkey in. We have a stack of old round foil deli trays in one cupboard and I'm trying to work out the logistics of somehow molding them into a makeshift roasting pan.

I haven't been to bed yet. I couldn't get any other queen to cover for me last night. The night before any major holiday is always a blockbuster night at gay clubs. Thousands of fags across the city fortifying themselves for long trips home, where they'll be met with awkward silences, stilted conversations, and cousins with whom they'd experimented decades ago.

I got home this morning at eight a.m., just in time to catch Katie and Al mispronounce the name of the first high school marching band in the Macy's Thanksgiving Day Parade. As I got out of my outfit, I watched the televised procession, taking place just a few blocks west of our penthouse. Growing up, I'd always dreamed that one day Larry Hagman and I would be perched high in the announcer booth overlooking the parade. Aqua wasn't born yet, but somehow I had an innate idea of who she'd become.

I envisioned Larry and I wearing our high-collared thick fur coats—his brown and mine white—as we read scripted jokes off the moniter and plugged our respective shows in between floats. He would reference my guest-starring role as Sue Ellen's best friend on *Dallas,* and I would teasingly rib him about why he hadn't yet appeared on my new hit sitcom *Extra! Extra!*—a workplace comedy with me as a wisecracking gossip columnist at a fast-paced New York newspaper. We would take turns "throwing it" to McClean Stevenson and Morgan Fairchild, who were covering the Hawaiian Thanksgiving Day Parade.

At the end of our coverage it would begin to snow softly behind us as I turned and announced:

"Larry, do you hear that? It sounds like sleigh bells!"

And he'd glance down at the teleprompter and say, "I hear it too, Josh . . . let's go in for a closer look, everybody . . . I think it may be . . ."

And then both of us together:

"Santa Claus has come to town! . . . From all of us here at our CBS family, to all of you out there, Merry Christmas!"

And we would laugh with condescending glee right through to the end of the scrolling credits.

. . .

I've finally successfully fashioned a sort of roasting pan out of tinfoil and a cookie sheet, and I am shoveling cornbread stuffing up the bird's ass.

"You need to curl your hand up tighter when you're fisting," Jack says, coming into the kitchen after his shower.

"This bird's plenty loose. One of your clients?" I reply.

"You do know that's going to take seven hours to cook?" Jack says.

"It's going in the cooker-thing in two seconds."

"It's called an oven," Jack says.

"It'll be fine," I say. "Open one of the champagne bottles; I want a little sip."

"Too early, Grandma," he says.

"I didn't ask for a martini, just champagne. It's like juice." He ignores me.

"Don't use this," he says, picking up the turkey baster I'd taken out of the drawer and put on the counter.

"Why?" I ask.

"It's for clients. You don't wanna know," he says, smiling, putting it back down on the counter.

In our newfound resolve to be a normal couple, Jack and I had invited twenty-nine assorted hookers, drag queens, club promoters, drug dealers, and Mr. Beefeater to our Old Fashioned Thanksgiving Family Dinner. Jack told them to show up anytime they wanted to, knowing that everyone's hours couldn't ever possibly synch up. At least I could safely assume none of them were morning people, so I still had a little time to prepare.

I don't really have that much to do. Jack went shopping yesterday and bought far more than we'd ever need. Huge plastic containers of green bean casserole, baked yams, Waldorf salad and cranberry sauce filled our refrigerator. Boxes of pies and cases of liquor were stacked outside on the balcony. I just need to peel about thirty pounds of potatoes and find enough pots to boil them in.

"I don't hear any champagne popping," I yell to Jack in the living room.

"What's the magic word?" he yells back.

"Cheers?" I offer. "DTs? Free blow job?"

The second I hear the cork pop, Jack's pager goes off.

"Fuck," I mutter under my breath.

I hear Jack take the phone into the bedroom. He doesn't like me to listen to him talk to clients on the phone. They expect him to talk rough, and he says I make him self-conscious. A minute later he walks into the kitchen and hands me the open champagne bottle.

"Gotta go piss on someone," he says.

"*Exactly* like Grandma's house," I mock. "How long?"

"Just a couple of hours."

"That's a lotta pee," I say.

"I'll be back before anyone gets here. Beep me if you want me to pick something up."

Fifteen people are here by the time he gets back. When he walks in the door, everyone lets out a cheer. I try to imagine how everyone envisions Jack and me as a couple. I like how we look. We're the Kennedys of Kinkiness. The Rockefellers of Wrongness. Maybe not the American Dream, but certainly a few people's American Fantasy.

The pressure of doing everything myself has led me to have a little more champagne than I'd planned. I'm pleasantly buzzed. Laura was helping me in the kitchen, since by the time she arrived I was having complications operating the potato peeler.

But everything is turning out close enough to perfect. One of the drag queens came as an Indian squaw and is carrying a load of airplane-sized little booze bottles in her papoose. She walks around asking the other guests if they "wantum sip fire-water." Old Mr. Beefeater comes in full palace guard regalia, and winds up being quite a help in the kitchen as well. The escorts come and go in a chorus of beeping. Thankfully the first thing they all do when they return from their calls is wash their hands.

The champagne also helps me keep my sunny disposition when I answer the door and Trey is standing outside. He holds out a bottle of wine, and I take his jacket. I glance at the label.

"Cheap," I say cheerfully, "how fitting."

"I didn't realize you were so picky about your binges," he replies, smirking.

I'm determined not to let him bug me. He hasn't been over since my little blow-up in the kitchen, and I'm pretty sure Jack isn't even going on non-"party" calls with him anymore. It's hard to imagine how someone could be a bad influence on a sadomasochistic hooker, but I consider Trey just that.

He heads into the living room and blends in with the rest of the guests.

Jack comes up behind me and squeezes me in a bear hug.

"Just like Grandma's?" I ask.

"Better," he says.

"The turkey should be done," I say. "Come help me."

Most of my extended family lives near Albany, New York. With my parents' divorce, my mother's remarriage, and our move to Wisconsin, we were the black sheep. Our holidays were small and intimate. Just the four of us. Simple celebrations that were used as occasions to teach my brother and me proper table manners and gave us the chance to sip wine, a skill I've long since perfected.

When I look around at the twenty-plus people filling every assorted table and chair in our apartment, I wonder if this large holiday gathering is something important I maybe missed out on. Every table has at least one person laughing, some person in the middle of a story, and someone else trying to interrupt with a joke. All of us are not at home, but couldn't be more at home.

I was fascinated by hidden picture books when I was little. The one that was my favorite had sweet nursery rhyme–type drawings hiding ghoulish and scary objects. A gingerbread house would have bats, witches, and skeletons hidden in its outlines. A rocking horse would be made up of spiders and ghosts. I would race, clutching my crayon, to identify and circle all of the craftily hidden evils.

Looking around our apartment, I see a hidden picture book drawing of a perfect Thanksgiving dinner tableau. A Norman Rockwell painting that, if you leaned in close, would discover is made up entirely of misfits.

Being drunk on champagne is a lot better than being drunk on vodka, I'm finding. I'm surprised I didn't learn about this sooner. Apparently you can teach an old drunk new tricks.

Our apartment has been full for almost ten hours. Nobody wants to leave, unless they get a call, in which case they come back immediately after.

"More champagne?" Mr. Beefeater asks me.

"Jack's not charging you for today, is he?" I ask him. There's no point in using Jack's work name, since everyone's been calling him Jack all night.

"No. Just here on my own."

"Well, Merry Christmas . . . Happy Thanksgiving . . . whatever the fuck," I say. "Sure, I'll take another one."

People have been going into my closet and bathroom and playing with my costumes and makeup. The result is that I'm surrounded by a dozen or so partial Aquas. I chastise them one by one for using the wrong lip liner with their lipstick, and holler out fat jokes when they can't cinch my corset.

I've found the perfect chair, which is an utmost consideration for the professional drunk. It's situated so that I remain in the middle of the action, it's comfortable enough so that slouching looks normal, and it's smack in the eyeline of whoever is bartending.

I'm happy. Jack has been catching my eye all night, and smiling that braggy smile that only couples hosting a successful party can smile. Everything's exactly right. The guests, the music, my blood alcohol level.

"Trey wants to party," Jack whispers, kneeling next to my chair.

"Surprise."

"I'll have them do it in the kitchen," he says.

"Where else?" I ask, as if everybody's kitchen doubles as a crack den.

"I won't even go in there."

"No worries. Do what you want to do," I say, feeling overly generous in my stupor. Jack gets up to give them the okay. "Wait," I say, flailing my drunken arm in his direction. He kneels back down.

"No public fucking afterward," I whisper. "There's too much food around."

Shortly afterward, I pass out and Jack disappears for five days.

22

W hat do you mean he won't talk to you?" Laura says, picking up a bad fifties oil painting and checking the price.

We're at the Chelsea Flea Market, and even though there are only two weeks till Christmas, the frigid weather has kept most people at home.

"He won't even look at me," I say.

"Do you talk to him?"

"Yeah. Of course. He just goes on doing whatever he's doing."

"Did he say where he was?" she asks.

"Did I not just say he won't talk to me?"

Jack only comes home for a few hours at a time, to change

clothes or make telephone calls behind the closed guest bedroom door.

"Not one word."

"Nothing."

"You need to get the hell out of there."

"I think he needs real help."

"Really? Ya *think?*" she says. "Look, Tutti, Jo's got in with the bad crowd and there's nothing you, Natalie, Blair, or Mrs. Garret's gonna be able to do to get her out of this jam. She's gotta love *herself* first."

"Shut up, Cousin Gerri," I say.

"Oh," she adds, picking up a vase, "and steal his cash on your way out. You owe me twenty."

For kicks, I'm lying on Jack's side of the bed tonight. Maybe I'll sleep better. It's been two and a half weeks since Thanksgiving eve, the last time Jack had been in this bed.

The master bedroom is farther away from the front door than the guest bedroom, so it's harder for me to hear when Jack comes and goes. Some mornings I get up and find a pile of his old clothes in the front hallway. It's the only way I know he's been there.

Even though it's high club season, I don't take any Aqua jobs. Three of the goldfish have died and I haven't even bothered to replace them. It's easier just to drink at home.

I feel like I'm in a tacky Lifetime movie, lying in bed

nights with an open bottle of Absolut in my hand. Sometimes I pretend I'm drunker than I am and stumble back and forth to the bathroom, theatrically delivering fictional monologues to nobody.

"You'll never get the kids," I slur to my reflection in the window, *"I'll tell them about your secret gay lover!"*

Or,

"I own fifty-one percent of this dump and I'll take you all down with me!"

It gives me something to do when I can't sleep.

The darkness of the bedroom brightens almost imperceptibly. There's light coming in underneath the door.

What's the point of getting up? The half dozen times I've been around when Jack drops by leave me more frustrated than when I'm alone. If he's high, he looks at me venomously. If he's crashing, he doesn't look at me at all. It's obvious that for some reason he blames me for his addiction.

But any chance to see him is still somehow better than nothing. Maybe this time he'll look up and I'll recognize something of the old Jack.

I get up quietly and peer out the door. The fluorescent overhead light in the kitchen is on. I can't stand the feeling of sneaking up on him in our own house.

"Hey, you," I say, standing in the kitchen doorway in my underwear. I expect no answer and get none.

Jack's setting up his equipment. I've never seen him, or anyone for that matter, actually smoking crack. I've smelled it. I've heard it. I've seen people prepare it and clean up after it. But I've always stayed away from the kitchen when anyone was actually hitting it.

It reminds me of watching my dad get his tobacco pipe ready. I would watch raptly. He would gather everything onto the table by his chair before he started. Pipe cleaner, tamper, tobacco pouch. The fermented apple smell and quietly intense look on his face were all part of the process. Process. I liked ritual. Tradition. Sinking into what you expect. Exactly what you already know is going to happen happens.

Jack has the same intensity as he gathers everything he needs around him. The glass pipette, the grill lighter, the Brillo pad. The rock.

He grips the Brillo pad by one corner and holds the lighter just underneath the opposite corner. It smokes as the invisible thin plastic coating burns away. I smell the beginning of a scent that's become as familiar to me as walking into a kitchen and smelling coffee.

He rips off a dime-sized piece of the stripped Brillo and carefully pushes it into the glass pipe. It's an old pipe and is coated with streaks of amber residue, which he scrapes off and collects on the little chunk of Brillo that he pushes back and forth with a piece of coat hanger.

He uses the coat hanger wire to position the Brillo chunk about a half inch away from the end of the pipe. It's the filter.

With the other end in his mouth, he picks up a tar-colored rock off a square of tinfoil on the counter, and puts it in the end with the Brillo.

He picks up the lighter. Flicks it.

He runs the flame up and down the length of the glass. A deep yellow smoke gathers and curls inside the pipe. He breathes in, still caressing the flame back and forth. The flame reaches his lips, then recedes back the length to his fingers. It's seductive. Like when he used to run his finger lightly up and down my back.

"Let me try," I say.

No response. The steady flame stays in constant motion.

"I just want to see."

I take a step toward him. My arm reaches up. I don't know if I'm reaching for the pipe or for him. I want to touch his skin. I want to breathe in what he breathes. The yellow swirl. I want to be the yellow swirl. I want him to breathe me in, be sent riding on oxygen molecules deep into his lungs. I want to travel through his body, seeing what makes him happy. Attaching myself to whatever place in him sparks to life on my arrival. His blood, his tissues, his muscles, I want to burrow inside the folds like a windblown dusting of snow, so that each time I melt away he seeks me out again.

There's no delineation between the pipe and the smoke and his body. It's all whole. I want in. I want him.

"Please," I say softly. "Let me try."

Without letting go of the pipe, he swings his hand holding

the lighter with incredible force, backhanding my face. My jaw pops.

The lighter swings back under the pipe. Undulating back and forth. Inhaling the curl as it rises from the tar. Exactly the same as before he hit me.

Only now he's staring at me. Hating me.

23

"Look. It's obvious I'm fucking things up," Jack says, sitting on a dining room chair facing me when I enter the front door. I set my bag down.

Yet another surprise turn of events. You never know what'll greet you when you enter apartment 42E. There's really no way to respond to what he just said. *Um, gee, I guess maybe you're right . . . I hadn't thought that your crack habit could be the cause of this mess my friends now refer to as my "living situation."* His pronouncement is as illuminating as walking into a room and announcing "I just walked into a room."

But what could be really interesting to talk about is why there are two fully decorated Christmas trees at either end of our living room and strings of lights framing all the windows and the balcony.

"Are you high?" I ask him.

"Not right now," he says.

"Congratulations."

He stares at me. I feel a little embarrassed for him.

"Were you high when you went and pillaged Whoville?" I ask, with a sweeping gesture toward all the decorations.

"Yes."

"Well, it looks nice."

More silence.

"Gonna stay a while?" I ask finally.

"Maybe."

"Well, I'm getting sick of taking out the trash myself every time."

"I want to stay. I have a plan."

"It's your place," I say.

He starts to cry. I've never seen him anywhere near tears. I've never seen his eyes even well up.

"Do you love me?" he asks.

The fear in his voice crumples me.

"Yes."

He puts his head in his hands.

More silence.

I hear a couple laughing out in the hallway. Our neighbors. I hear their keys jingling, and their deadbolt click, and their voices recede as they step inside their apartment. All these people in boxes.

"Yes." I say it again.

"I need to be able to smoke here for a little while," he says, looking up.

"It doesn't seem like you've been holding back," I respond.

"I just want to be able to smoke here without you not liking me," he continues. "I just want it to be us . . . I want to get through Christmas . . . I just want you for Christmas."

His voice catches on nearly every word.

"I can't stop now . . . not right now . . . New Year's . . . I'm going to Baja Peninsula for a month . . . by myself . . . New Year's. I bought a ticket."

Cactuses.

"I just need to go to the desert . . . when I get back I'll be clean . . . we'll start over," he continues. He's having a hard time getting any words out. "I just want to give you your Christmas."

He's shuddering.

"Please let me smoke here till Christmas . . . I want to give you Christmas."

The difficult part of having been raised by popular culture is that when confronted with a melodramatic situation, I flip through hundreds of similar sitcom plots and movies of the week to try to find an appropriate response. Is it time for the "very special tough love" episode? Or the weepy abused codependent spouse speech. Unfortunately, there's no commercial break time to sort it out.

"You can smoke here. I'd rather that you did," I say.

Jack's relief is physically visible. I slide down next to him and take his head in my hands. His lips are burnt, and his eyes so deep set I'm not sure if they're ringed in black or have receded so much that they're in shadow.

"I can't stop it here. I have to get away."

"I know," I say. "It won't work here."

Jack turns away from me and looks out the window. The winter sun is setting, sending shadows of skyscrapers halfway across the island and out into the East River. There's a moment as the sky grows dark each day over Manhattan when the light from the streets and buildings is equal in luminescence to the fading light in the sky. It's a fleeting second, but all at once everything has the same dull sickly orange glow. It's the city after the sun's cleansing brightness and before the evening's artificial sparkle. It's simply a city without any makeup.

"Do you like the trees?" Jack asks after he's composed himself.

"They're beautiful. How did you do all that?"

"That's what happens when you get an idea while you're tweaking," he says, sheepishly.

They are beautiful. Positioned at either side of the room, they reflect in the wide expanse of windows against the dark glass, making the room appear as if there were dozens of sparkling trees.

The truth is, there's no movie of the week about a drunk drag queen and a crackhead hooker in love. There never has been. It's not the kind of thing people would care about. People would flip right by the channel, either unbelieving or uncaring. Who's the good guy? Who's the bad guy? Aren't they both bad? If they didn't get what they deserved by the first commercial, it'd be on to the breast cancer movie.

"You'll be gone for a month?" I say finally, breaking the awkward silence.

"Yeah. I'm just going into the desert."

"Will you call me?"

"No."

"Can I stay here?"

"Of course. When I get back, we'll figure things out."

I'm petrified of losing him. At Thanksgiving someone asked Jack how long we'd been in our relationship. He'd said we weren't in a relationship. We were in a conspiracy. We were. Are. As damaged as we may be, I know we have more going for us than any couple in any box in this city. If we can find a way not to damage ourselves beyond repair, we will make others doubt themselves as much as they doubt us.

"When you get back, will you want me?" I ask.

"I want to want you."

It's not the answer I want to hear, but getting any type of answer is an improvement over the last few weeks.

I look out over Jack's head in my arms at the skyline beginning to light up, sparkle by sparkle. It's impossible to conspire by oneself. Secrets that reside only in the mind of one person aren't really secrets. They're unspoken fears. It takes two to conspire. I kiss his hair, soft and brown, and breathe in deeply, smelling pine sap, and honey, and desert, and worry. I may be foolish, but I'm not stupid. I've learned a lot from Jack. Mostly to take things as they are at the moment they happen. And right now, this moment, I'm happy to be invited back into the conspiracy. I want to memorize his presence because I know what it's like when it's gone. I want to never not know this moment.

24

"Isn't it Boxing Day or something? Shouldn't you be home waiting on your butler?" I ask Houdini as I open the door to greet him.

"Merry Christmas, little tart," he says, smiling, handing me a large white shopping bag. "It's for Lady Aqua."

He stomps the snow off his boots in the hallway and steps inside. He unwraps his plum-colored cashmere scarf and pulls off his black leather gloves.

"Here," I say, taking them. "Give me your coat."

"How was your Christmas?"

"Brilliant," I say.

It was. If there's ever a good time to have a boyfriend high on crack, my vote would be for the holidays. Especially if it's a boyfriend who feels a little bit guilty for ruining your life.

One night when Jack and I first began dating, deep in the swamp of summer, I'd gone on and on in a drunken ramble about how much I was looking forward to my first New York Christmas. It had always been a fantasy of mine to move into the Plaza right after the Thanksgiving Day Parade and take part in every bit of New York Christmas right up until the Times Square ball drop.

Jack remembered my mentioning this, and for the week leading up to Christmas, Jack had a different activity planned each night. Skating in Central Park. Caroling at Washington Square Park. A hired car took us from department store to department store so that we could see every holiday window display in the city. On Christmas Eve, he took me to St. Thomas's Episcopal Church for midnight mass. When the service was over and the congregation spilled out onto Fifth Avenue, the bells of St. Patrick's, St. Mark's, and St. Thomas's were pealing simultaneously. If it had started to snow softly, I would have been twirling, arms out, down the middle of Fifth Avenue.

I haven't even been in New York a year yet, but I've seen enough for cynicism to take a strong root. Jack's tour of the New York holiday season broke all that down, and gave me the wide-eyed innocent Christmas story I'd always dreamed of.

I'd taken a break from Aqua, turning down all holiday gigs. Jack and I went to bed together each night, though just as I would relax, that moment when I felt like I was falling, Jack would slide out of bed and head into the kitchen. I would hear the cupboards and the lighter flicking, and it was unexpectedly comforting. Sometimes I would wake up in the middle of the night and hear muffled voices with him. Trey, Marcus—another escort I'd met, strange voices. It didn't matter. The conspiracy had survived. The voices, and the auditory ritual rhythm of his getting high, relaxed me. Habit. Tradition. Home.

Occasionally I'd get up and join him, or them, in the kitchen just long enough to pour a juice glass of vodka. They were high, and sometimes nervous around me. But then they'd follow Jack's cues and relax. We'd smile and exchange pleasantries, and I'd head back to the bedroom to watch *Barney Miller,* or *Cheers,* or an infomercial.

On Christmas Eve Jack slept with me the entire night. We woke curled together exactly as we'd fallen asleep.

"Stay in the bedroom a second," he said. "I'll come get you."

I turned on the TV and watched the *Today* show. I was offended momentarily when Katie and Matt were absent, replaced by some second-tier substitute local anchors roped in by an unfair holiday clause in their contracts. These two people are in our homes every morning of the year and then abandon us on the most important holiday? It didn't seem fair. I think of how I'm not at home, my first Christmas away in twenty-six years. My mother told me that of course she would miss me, but that she was happy I had Jack and would be thinking of us both.

"Come out," Jack called, cheerfully. When I emerged into the living room, I noticed how bleak Jack looked. Not having gotten high the night before, his body, his entire being radiated a sort of wired exhaustion. An empty nervousness that made me hollow just looking at him. But his smile was genuine as he gestured at the pile of presents he'd assembled under both trees.

"Oh, Jack," I said, "it's too much."

"Come here. You need to open them in order."

He explained to me that he'd been collecting gifts since last summer when I'd told him about my New York Christmas dream. One gift from each section of town.

He handed me a lumpy package wrapped in red foil decorated with gold dragons.

"From Chinatown," he said. I carefully peeled back the tape from the folded ends, and an assortment of candy and intricate small toys tumbled from inside with colorful Chinese characters on their packages.

Dozens of packages followed. Versace jeans from Fifth Avenue. A photography book from SoHo. Silly gay toys and skimpy decorative underwear from Christopher Street. Museum passes from the Upper East Side. Opera tickets from Lincoln Center. There were endless gifts. And a card with each one, with drawings and poems about the neighborhood of origin and its significance to us.

As I reached the end, I couldn't bear the thought of it stopping any more than any child around the world opening presents at the same moment. Each gift was tangible proof that I was loved and thought about and cherished and worthy. Christmas is not about giving; it's about feeling deserving, the warm innate joy of knowing good things will come to you, that forever someone will provide.

Jack was lost in my world with me as I opened the presents. He opened my presents to him and professed the appropriate amount of surprise and excitement, but there was nothing I could give him to equal my satisfaction with his gift of my first New York Christmas. Neither of us were prepared for the last gift to be opened, for the last ribbon to be undone, and paper to be discarded. Neither of us wanted to face the enormity of the vacuum of what was left in front of us, Christmas afternoon and beyond, the empty uncertain distance between this scripted moment and the rest of our lives. This was the end of our plans, the edge of our cliff. But we had made a promise to each other, and we kept it. Perhaps for the first time ever. It felt good.

When we were finished, we sat crosslegged in our underwear on the cold parquet floor. No noise came from above or below or beyond the windows. The city on holiday—frozen, suspended, empty. The skyline outside, clear in the gray winter sun, frozen like a backdrop in an empty theater. It was easy to believe we were the only two who had been celebrating at all, and as soon as we were done, the crushing void moved in.

. . .

"Aren't you going to look inside?" Houdini asks me. I'd been lost in my Christmas memories.

I open the white bag and peer inside.

"Lush! My favorite store!" I say.

"I know, it's all you talked about last time," Houdini replies.

Inside the bag were dozens of soaps and creams and bath bars from my favorite London store.

"Thank you," I say, reaching one arm around him and kissing him on the cheek. Houdini briefly stiffens against my embrace, but then reaches an awkward arm around to pat my back.

"Where's Aidan?"

"He knows you're coming, right?"

"I phoned him on the way to Heathrow."

"He'll be back shortly," I say, without any idea if I'm telling the truth. After a brave attempt at spending Christmas afternoon together reading the paper, watching TV, and dancing around to the new Celia Cruz CD I'd bought him, he'd slipped away as the sun set and I hadn't seen or heard from him since. I guess I owe him one last binge.

"Come on in. Can I get you something to drink or eat?" I ask. "I don't think we have any coke yet. I haven't seen any around."

"No worries. I'll just wait for Aidan."

We spend the next half hour talking about his Christmas, and his daughters' presents, about his business, and European politics. Jack's absence is beginning to grow uncomfortable, and I wonder if I should start bossing Houdini around and telling him to undress. The guy's come a long way to be stood up. I'm trying to remember where Jack stores the restraints when I hear his keys in the lock.

I jump up to meet him at the door.

"Houdini's here," I whisper.

Jack looks terrible. The same white T-shirt he wore as he left the apartment Christmas evening is now filthy with yellow stains under his arms. His lips are chapped and bleeding, and the acrid stench of burnt cocaine and butane rises from his skin. He walks past me into the living room, his hollow eyes having never looked directly at me.

"Hey," he says to Houdini in his lower than normal escort voice. "Take your fucking clothes off, pussy."

I step softly into the bedroom, suddenly feeling like an audience member who'd accidentally found himself onstage after coming back from the bathroom.

The rest of the afternoon passes outside my closed bedroom door, marked by Jack's gruff barking of commands, and Houdini's high-pitched pleading and muffled struggling. I flip through the channels, trying to find anything that will hold my attention more than a few minutes. By sunset I've decided to put on Aqua and go out, even though I have no paying gig and can't find anyone left after the city's holiday exodus to go out with me.

"See you boys later," I call, traipsing through the living room wearing a red vinyl catsuit with white fur trim around the collars and cuffs, so skintight that I begin to sweat before reaching the door.

Jack doesn't even look up from the magazine he's reading, and Houdini is too afraid to break his victimized character to respond. But I see through the restraints and the bound naked fatty flesh, and wink at the husband/father/executive lying on his side on the floor. His eyes dart back and forth between me and Jack, who's so exhausted he's paying almost no attention to Houdini at all. Finally Houdini winks back at me, piercing through the makeup and wig, and making me glow underneath it all. We're double agents. Conspirators against the conspiracy.

I'm amazingly undrunk when I return at the remarkably responsible hour of four in the morning. The clubs were fairly empty, and the post-holiday vibe was relaxed and relieved, the few people that chose to remain in the city coagulating into a sort of adoptive family unit.

Jack stands in the kitchen, still in his stained T-shirt, pulling together the ingredients he'd need for the next day's high.

"Hello again," I say.

No response.

"Hey," I say.

He continues measuring out the rubbing alcohol into a small juice glass.

"Fucker. Look at me."

He does.

"What the fuck? You asked if you could smoke here; I said yes. You asked if I'd not hate you for it, and I haven't. You said we'd just ride it out till New Year's, then you'd disappear and clean up, and I was cool. The least you can do, the very fucking *least* you can do, is be civil."

I am not getting back on the roller coaster.

"Sorry," he says, turning back to his methodical task. "You're right."

I feel completely unvindicated, and a little ashamed at my outburst. He gave up too easily. Sighing, I put my bag down next to the sink and lean against the counter, watching him work.

"When do you leave for Baja?" I ask, breaking the silence.

"New Year's Eve Day," he says.

Two days away. I watch him pouring and measuring. Tipping the ingredients into one of the spoons from the set my mother gave me when I went to college.

I loosen my corset and take a deep breath. The kitchen is full of his musky unshowered scent and I want to bury my face in his back and hold his arms at his side to stop him from what he's doing. Instead, I take another breath and hold it as long as I can.

"Jack?" I say after a few minutes have passed.

No answer.

"Do you want me here when you get back?"

I hadn't planned on asking this. Not out loud, at least. I'd been asking it in my head constantly since the afternoon he'd revealed his plan to me, but I'd never meant to say it out loud.

"I don't know," he says finally. "Do you want to be here?"

Yes! I want to scream, *Yes! Yes! Yes! Yes! Yes!* over and over again until my throat dries out and bleeds out over my tongue. *Yes, I want you to come back and be clear, and bright, and stunningly perfect. I want you to tell me what I'm doing too much of, and not enough of, and convince me that we'll go down for all eternity as the perfect couple. This isn't all about you.*

Instead, I tell him that I don't know where I'll be when he returns. It's more of a show of bravado than any version of truth. I know I'll be here. Unlike Jack, no matter how beneficial a disappearing act might be for me, I could never tear myself away from a show in progress. Even when the plot's tragic ending is apparent to the entire audience. Perhaps there's a deus ex machina that will lower from the ceiling and turn the whole debacle back into a romantic comedy. Never can tell. Paid the full ticket price, might as well stay.

Plus, I have no where else to go.

I head into the bathroom to undress and shower, reemerging wearing Jack's thick white robe. He's finished in the kitchen, and I wipe up his mess.

"I'm going to bed," I say, standing in the living room doorway. I'd forgotten Houdini was still here. He's lying still next to the futon, arched backward in his normal hands-and-ankles-tied-behind-his-back position. Very still. His eyes are closed.

Jack's flipping through *The New Yorker* on the couch.

"What's up with Houdini?" I ask him.

"Resting, I guess."

I step softly over to him and crouch down. His pale skin is slick with sweat, the sparse graying hair on his chest speckled with dirt and dustballs from the floor. In his struggles, he's flipped his water bowl over completely and knocked it six feet away under the corner end table. The two-day stubble over his lip is coated with cocaine, and his right nostril had been bleeding at some point earlier and now is clogged with a large black scab. The trail of dried blood runs in a cracked black streak down the right side of his face before dissipating in a dark wide smear where his cheek rubs against the floor.

His stomach and back are bruised an angry red from Jack's kicks.

"How long has he been out?" I ask, turning to Jack.

"I don't know. A couple of hours," he replies, never looking up from his magazine.

I've never seen Houdini sleep before. The whole point of his visits is to get as high as he can get so he gets his money's worth of struggling against the restraints. I stare at his sleeping face. I want to put one of the throw pillows from the couch underneath his head, but instinctively realize this small comfort would ruin his trip. I imagine his bed at home, with six-hundred thread count cotton sheets, multiple lush down pillows, his soft heavy body sinking into his luxury mattress, his thick pink arm curled over his graying sleeping wife.

"I'm not sure he's breathing," I say.

"He's fine. We're just taking a break," Jack says.

"Really. Come check it out."

I softly place my palm against his chest. His skin is sticky with sweat and oil and hot to the touch. But instead of the soft even rising and falling of sleep, his chest almost imperceptibly vibrates, shaking and panting with silent breaths almost too shallow to feel. It seems more like a seizure than actual breathing.

Jack's standing over me now.

I reach behind him and try to wriggle my fingers under his wrist restraint. I don't know how to check for a pulse; I'm copying what I've seen on television. It's not working; all I feel is the pulse of my own fingers pressing my fingernails back against the leather cuffs.

I lay my ear on his chest. It takes a moment to distinguish a rapid tapping from the rest of the whooshing and humming symphony coming inside. It sounds nothing like any heartbeat I've heard, any familiar restive two-beat liquid rhythm. I hear someone nervously drumming their long fingernails on his rib cage.

Jack stares down at me.

"I think he's OD'd," I say. "Call someone."

Jack continues to stare.

At some point in the very near future one of us is going to have to act. Until one of us actually does something, it could be argued that nothing's actually happened. This is what I'm thinking instead of thinking about what to do next.

"Go into the bedroom," Jack says calmly.

Because a part of me thinks that he's come up with a way to undo this moment, and another part of me believes that this is his mess not mine, I do exactly as he says.

Less than five minutes later, as I'm blankly staring at Mr. Rogers and King Friday discussing "hurt feelings," Jack yells for me to come back into the living room.

Houdini is unbound and dressed in the rumpled lavender dress shirt and navy slacks he showed up in yesterday. He's slumped, still unconscious, in the corner of the futon. Jack's struggling to wedge Houdini's impossibly shiny loafers onto his bare feet. Without socks, his clammy skin won't wedge inside.

"Find his underwear and socks and stuff them into his bag—it's in the closet by the door," Jack says in an even tone, still struggling with the reluctant shoes.

His underwear is under a dining room chair, and I find one sock kicked behind one of Jack's Mexican tin statues. I can't find the other one.

"Ready?" Jack asks, pushing Houdini forward enough to wrestle his Burberry trench coat back onto him.

"For what?" I ask.

He doesn't answer me. He finishes struggling with the coat and walks over to the house phone in the front hallway.

"Hola," I hear him say before continuing in quick Spanish. To Pedro, I suppose. I decipher nothing except for the words Jack doesn't know Spanish for: "elevator," "garbage day," and "wrapping paper." He wraps up his conversation with a clipped laugh and "Gracías, adios."

"Let's go," he says. "You carry the bag."

Jack stands in front of Houdini, bends over and hugs his body to him, lifting slowly and awkwardly till Houdini's on his feet, slumping his full weight into Jack's hug. I think of the performers in the subway who dance and twirl with life-sized stuffed dummies attached to their feet. Only Jack's partner is close to two hundred and fifty pounds and semiconscious. Jack struggles with his grip, pulling himself around behind Houdini, dragging him toward the door.

"Open the door."

I grab Houdini's leather duffel bag and sling it over my shoulder.

Jack looks both directions down the outside hallway before dragging his slumping cargo down toward the service elevator. Pedro has sent it up to our floor already, and the doors open as soon as I hit the button. I get in, following Jack and Houdini.

The elevator descends directly to the garbage bay that opens onto Seventy-eighth Street. On our trip down, Jack rearranges himself again so that Houdini leans onto Jack's right shoulder. Now he simply looks like a businessman who's been out drinking all night.

Houdini starts mumbling. "Out. Out. Go out. On the steps." His voice is thick. He says something like "house," only drawing out the syllable: "howwwwwse."

Jack slips his arm around Houdini's thick waist.

Out on the street, the frigid wind swirls up my T-shirt, billowing it out and freezing my skin into prickly goosebumps. I hug Houdini's bag close to my chest for more protection. Jack turns left. The river.

I jog around them, Jack's tricep bulging and taut as he lifts Houdini's weight high enough that his feet barely drag on the ground. It must be around five thirty in the morning and the streets are blessedly barren. Houdini's mumbling more. It sounds like questions now. Or a sing-songing attempt at starting up a conversation. I pick up my pace, trying to gain at least half a block between us, trying to separate myself from the strange duo behind me. Though with no one else on the sidewalk, anyone passing immediately would connect the stumbling pair with the suspiciously underdressed guy tightly hugging a suitcase to his body.

This is not my mess. This is not my mess, I repeat over and over in my head, trying to crowd out any other thoughts. *This is not my mess.*

I reach a small park at the end of Seventy-second Street that overlooks the East River, but don't turn around until I approach the metal railing overlooking the inky rolling currents of the water. Jack's still a hundred yards behind, moving slower as his strength, built up over years of rigorous workouts, begins to reach the end of its capabilities. His solid muscle, refined and maintained to titillate for profit, now lumbers straining, taut, stiltingly toward the park. Houdini's head is flopped down against his chest, bouncing along with Jack's heavy steps, nodding a continuous "yes" as he glides down the sidewalk.

Not my mess, I continue as the pair approaches.

I try not to look at them, and instead absently scrutinize the bag I'm holding. Down next to the river, the wind is violent, coming from no particular direction. The early morning traffic on the FDR is picking up, roaring past on the highway above and behind me. If it had been summer, the path along the river would be just starting to fill up with morning joggers and bicyclists and rollerbladers. Instead, the über-healthy people were at their gyms, sweating away on treadmills and stairmasters, earphones plugged into CNBC. The promenade is completely empty as far as is visible in both directions.

I continue to stare at Houdini's bag. His luggage tag flutters in the wind before flipping up and coming to rest in the crook of my arm.

Donald Ranthrowe
6–17 Pembridge Garden
Notting Hill, London W2 4DU
UK
020 7229 5396

"Donald."

Houdini's name is Donald. Don. I wouldn't have guessed that name. But it doesn't not fit. His wife's name is Elizabeth, and his daughters are Dawn, Grace, and Penny. This I knew from our conversations. But I didn't know "Don." Is it Monday? What time is it in England? Are his daughters sitting in their school? Taking a test? Passing notes to boys? Is Elizabeth at lunch with her friends? Shopping for dinner tonight? Two dinners, one for the girls and a late one for when Don gets home, tired and grumpy and complaining about the New York office? What do they look like? I wish I had asked Houdini to show me pictures. Don.

From behind me I hear a cracking pop as Jack slides Houdini down onto a bench.

"Give me the bag."

Houdini's slouched into the corner of the green wooden bench, his head lolling to the side, eyes, open now, staring at nothing, still mumbling a conversation full of unintelligible questions that no one can answer. Jack places the bag on Houdini's lap and pulls his arms on top of it to hold it in place. I place my hands on top of his. They're hot and clammy and fidgety. His eyes are open, but they're looking through me.

Jack's halfway back up Seventy-eighth Street by the time I tear myself away from the babbling Houdini. Suddenly afraid of being spotted, I take off after him, half-jogging. He stays far ahead, taking long, determined strides, and by the time I reach the apartment, he's already in the bathroom, door closed, taking a shower. I stand shivering in the foyer, reading and rereading the headline on the *Times* that was delivered outside our door during the fifteen minutes we were gone.

Houdini's Christmas gifts for me are still sitting in the white bag at my feet.

This is not my mess. I try to convince myself. I think of Houdini reading me a recipe for Sticky Toffee Pudding from the paper last fall, pointing out the differences between what he was reading and how his grandmother used to make it. I promised him I would try to make it for him one visit, and he laughed. *I'd rather have gingerbread and custard sauce if you're taking orders,* he'd said. No problem, I told him. What will he think when he comes to? Will he remember I was there? My stomach churns, emptied of vodka, and filling with bile. I need to eat something. I need a bowl of cereal, so I can get ready for school, sit on the bus reading my homework, putting the finishing touches on my extra-credit report. I need to make people happy instead of continually making messes.

I change into jeans and a sweatshirt, grab a jacket from the hall closet, and head back outside.

Pedro waves to me in the lobby from behind his station.

"Busy morning!" he calls out, smiling.

"Lots to do," I say back, trying not to seem like I'm rushing as the automatic doors slide open.

In the five minutes I've been upstairs, the city has awoken. People dot the sidewalks, clutching steaming take-out cups of coffee and briefcases. A group of four women in long black wool coats peer from the bus shelter, looking down York for the next bus. I duck into the phone booth next to the shelter.

"Nine-one-one. What's your emergency?"

"There's a man on a bench by the East River—around Seventy-eighth Street. I think he's sick." I hang up.

The cold front that blew in the night before hangs like a purple frozen wall over Brooklyn. The sun, pale and pink, rises from behind it, feebly stabbing at the thin winter cold. The steam from a factory across the river puffs out into the sky, frozen white, not blowing, just expanding high into the air. I'm just walking. Just out walking. Going to walk along the river.

Far down York a siren begins its warbling crescendo. A staccato chirping when it reaches each intersection.

Hands in my pockets. Just walking.

25

I'm surprised at how not empty the apartment feels once
Jack's gone. Wednesday morning there was a large backpack
sitting by the front door, stuffed with guidebooks, PowerBars,
jeans, T-shirts, and a portable CD player. By Wednesday
night it was gone.

In his absence, I realize that since Thanksgiving, Jack's
been like a special guest star on our own show—like a main
character who's had a contract dispute with the network and
consequently only shows up sporadically for filming.

Aqua, of course, gamely carries on with the show by herself.

I spend the rest of the holiday season working each night,
surprising myself with, if not sobriety, something far short of

the walking cirrhosis poster child I usually am. During the day, I attack my advertising assignments with a ferocity that frightens Laura.

The apartment feels more like a waiting room than home during the short visits I make there between jobs. Mostly I just nap in the few hours I have before heading to a club or back to the agency. Every minor issue dealing with the future turns into a big black question mark on the front of my brain. There's only three more rolls of toilet paper. Do I bother buying more? What's the point of changing the burnt-out lightbulb in the front hall? Should I buy a whole gallon of milk, or will I be gone before it is?

Tonight, at Cheetah, we had a private party for the owner's birthday. It wrapped up early and I'm home in bed before three. The prospect of five hours of uninterrupted sleep before having to be at the agency wraps me in a luxury of contentment that ironically makes it hard to doze off. I'm sitting reading magazines and watching Nick at Night reruns when I hear a key clicking in a lock.

I can't think of any of our neighbors who have ever been out this late. Before I even hear the thump of his backpack hitting the floor, I know it must be Jack.

The light flicks on in the hallway.

Why is he back?

Jack passes through the bedroom into the master bath without acknowledging me. He turns on the water in the sink and starts splashing it on his face. I turn on my side and watch him. He scoops up handfuls of water and presses them tightly to his face, holding his hands there long after all the water has slid back down into the sink. Rubbing his fingers into his closed eyes and breathing deeply before returning his hands under the running water and starting again.

He's so thin. When did he get so thin?

I get up to pee.

Standing at the toilet next to him, I pull down the waistband of my underwear and relax a steady stream into the bowl.

"Hello again," he says in a defeated monotone.

"Trouble finding the airport?" I ask him.

"Didn't get there. Got a call on the way. Party call," he says through closed hands over his face.

I finish peeing and reach for the flusher. With my other hand I lower the toilet lid and sit down. He goes on splashing his face.

"And now?" I ask. I'm hoping he has more answers about what's coming next than I've been able to conceive of in the last week.

"I'll change my ticket and see if I can get out of here tomorrow or the next day."

"Take a day to recover; you look like shit."

He stands with his fingers pulling down his lower eyelids. He stands there a moment, transfixed at the bloodshot eyes staring back at him. If he hooked his thumbs in his lips and pulled them up into a smile, he'd be making the face my dad used to make to force me to laugh.

"I think when I come back you shouldn't be here," he says finally.

After months of wrestling with vagueness, I am as relieved at finally knowing what I am expected to do as I am petrified of having to do it. Once again I am reminded that I am, and have always been, only what is expected of me. And at least now I have a purpose again, a concrete goal, even if it's only to not be around anymore.

I reach over and turn the faucet off. There's a sudden spasm of pain in my gut.

"I have to use the bathroom," I say.

"You just did."

"I have to *use* the bathroom," I say euphemistically. He dries his face on a towel and wordlessly exits the bathroom. I get up, lift the lid on the toilet, and sit down again on the seat.

The pain in my stomach is growing stronger, and when I thump it with my fingers, it's rock hard and echoes. Just gas, probably. The combination of alcohol and corset does strange things sometimes.

The pain intensifies, but I somehow feel relaxed. At least the sharp stabbing is a feeling. An honest-to-God physical sensation. I've been on autopilot for months. I haven't allowed myself any joy or pain or empathy with myself, Jack, or anyone else. I've just set the cruise control to "autism" and sped on down the road.

Now at least I have physical pain to work with.

Eventually, my bowels relax, and I go. Looking down into

the water I'm met with the most alien excretion I've ever seen. It's actually glowing. Sparkling.

For the party at Cheetah I'd tried a new lipstick process. I'd painted my lips with a sticky gloss, then patted raw iridescent pigment directly onto them. The resulting effect was stunning, but practicality-wise the new look was a disaster. I had to pretty much reapply from scratch every time I finished a vodka, having ingested more and more of the pigment with every sip.

I open the door to the bathroom.

"Hey, come in here and look at this," I call to Jack.

He appears around the corner, probably relieved that for the first time in months we might have something to talk about other than crack, booze, and our faltering relationship.

"Look," I say, pointing down into the toilet at the thin sparkling ribbon lazily floating in slow laps around bowl.

He stares.

"You're shitting glitter," he says.

Indeed I was.

I flush and we head into the kitchen to order our breakfast from the deli for probably the last time. Just like normal.

26

This is what I am dreaming:

It is Easter morning and my older brother and I are standing in our frozen front yard in Wisconsin. He is about ten, and I am about eight. The crab apple tree in front of us is still a good month away from budding, and the scaly black branches are decorated with hanging drugstore plastic eggs of various technocolors, vivid against the dark branches and pale blue early spring sky.

"Take one down for me," I ask him.

"Leave 'em be."

I want all of them so badly.

A frigid breeze picks up and sends the eggs swaying and clicking against the stiff branches.

"Take one down!" I ask him again.

He reaches up and pulls one off for me. As he hands it over, I realize that it wasn't hanging by threads, it was hanging by a fuse, which sparked to life the moment it separated from the tree. Simultaneously, all the eggs release from the branches and fall to the ground, their fuses crackling and sputtering.

I rush to gather them all in my arms, ignoring the burning fuses.

"Time to go, buddy," Rick says.

"Hang on, not yet." I'm furiously grabbing at the slippery plastic eggs with frozen fingers.

"Time to go," he says again.

When I wake up, the door to the balcony is open, and the freezing wind has blown the bedcovers completely off my feet.

And Jack's standing over me with my good Wüsthof chef's knife in his hand.

27

Every vodka has a story. And the ending is reliably rosier than the beginning.

The one I hold in my hand now arguably could be easily underestimated. A middle child in an expanding brood, it knows the stolid nonglamourous role it's expected to play in the evening. It promises neither the first flush of disassociation that arrived hours earlier on the backs of its older siblings, nor the final fade-out of memory and reality to be delivered by an icy baby hours from now.

I pretend to myself that I use alcohol to escape, but actually, as time goes by, I'm discovering that drinking sends me home. Here at Tunnel, I am at home. I know that each hidden corner of each darkened room holds a comfortable promise of sex, or fan attention, or simple companionship. And like a gracious host, the club offers me a constant stream of liver-darkening booze, which, if turned down, would be impolitic. The fantastical world I craved as a child is now all too familiar. And habitual.

Jack has left again, and this time I instinctively feel that he's been successful in escaping the city for the desert. I haven't yet thought logistically about leaving the apartment. There's a blissful empty month ahead before I have to fulfill my orders of disappearing, a whole blissful month so void of potential surprises that it's surprising in itself.

"What are their names?" a girl in a too-tight Spandex top and a frozen narcotic grin asks me, tapping on my breasts.

"Vodka and Tonic," I reply. She's probably only nineteen or twenty and makes this trip into the city from New Jersey every Saturday night with her high school friends, who, one by one, will drop out of the gang and begin their lives of husbands, and jobs, and children. She'll hang on, high and giggly, far past the age where she attempts to logically defend her partying, and her group of friends will shift and rearrange themselves until it's just her and a handful of gay twenty-somethings who pretend to adore her but laugh at her outfits when she heads to the bar for another drink. Her gay friends will balance boyfriends, and careers, and increasingly infrequent nights out, while she will find more and more outlandish costumes and personas to eek out some sort of attention from the world that has passed her by.

"Speaking of," I continue, "they could use a little company." I hold up my empty glass.

Sometimes when I send someone off to buy a drink for me, they disappear permanently into the crush of partyers. But she'll be back, most likely with a double, and several of her friends in tow. "Look who likes me," her eager grin will broadcast to her friends, and once I have the drink safely in hand, I'll turn to her cute friends and ignore her.

It's an average night. An average crowd, and average music. On nights like this, all I want is to find a seat, drink, and wait for the sealed envelope with my three hundred and fifty dollars to be dropped into my bag. But tonight I feel industrious. I will be a valued member of this clubbing society. Ask not what the night can do for Aqua, but what Aqua can do for the night.

The group of boys surrounding me is beginning to bore me. Each trying to come up with a pithier line than the last, trying to make the drag queen laugh. I drain the drink that mysteriously appeared in my hand, and thank the group without having any idea if it came from them or not.

Dance. I think I should dance. Mingle a little on the main floor. Earn my keep. On my way to the dance floor I'm stopped every few feet or so by someone who wants to stare at the fish, or tell me about how they went in drag last Halloween and looked *just like* a real woman.

Finally turning the corner onto the main dance floor is like turning the corner into a seizure. Strobe lights intermittently light up the crowd with the beat of the music and then plunge the room into complete darkness between the heavy thumps. The force of the music overpowers the muscles; it's impossible to walk, or breathe, or even blink out of rhythm.

The vodka works better here. I inhale lungfuls of air, a complex mixture of sweat, smoke, and the chemical traces of the smoke machine that faintly remind me of our kitchen crack lab. Moving now. The powerful bass replaces my heartbeat, and I move into a group of boys near the wall. Their faces light up with my arrival, like a group of five-year-olds when the birthday clown shows up.

Encompassing me in a circle, I pretend not to look at any one of them, alternating between gazing at my feet as they move back and forth and tilting my head back and swinging my hair to the rhythm. The movement makes me feel drunker than sitting, and my arms are growing numb.

I look down at my fish, one fish in each breast, lazily sloshing back and forth. They are face to face, staring at each other. What are they thinking? So close to each other, but separated by a divide impossible to breach. Or do they each think the other is a mirror? Or do they even think?

One of the boys grabs my waist and spins me toward him. He's cuter than the others. Definitely the leader of the pack. Of course I think he looks a little like Jack. He takes off his shirt, never stopping his swaying, and his flat, rock-hard stomach arches and grinds. He puts his hands on my waist and pulls me to him. When our middles meet, his hands reach around and grab my ass, moving me with him, perfectly together, liquid with the sound. His friends widen their circle as we dance together, all of us lip-synching the monotonous sultry chorus of the song to each other.

And suddenly I am there. I am where I try to find every time I put on the wig, and the makeup, and the goddamned heels, and walk out my door. Right this moment I am who I am supposed to be. My *raison d'être* fulfilled. I exist to hunt down one person, and separate him from the rest of the mob. The rest of the herd who bounce, dully, hypnotically, along with everyone else. A great mass of nothing swaying and nodding blankly at each other.

I take this one person and save him from the crowd. Whether they want it or not, they find themselves decorated by me, rescued from the expected. And I plead, silently and invisibly, that he will want to save me back.

It's too hot now. I should have had some water before going on to the floor, but I don't want to leave. This is the one I'm taking home tonight. He's not ready yet, but I'm not going to leave him for something as silly as water and risk breaking the connection. One of his friends taps me on the shoulder. I turn to him and he puts a finger on the side of his nostril and sniffs, one eyebrow raised questioningly.

Why not. A little coke will keep me going as well as water could. I nod yes, still dancing, and the boy dumps a dime-sized clump of white powder from a plastic packet onto the first knuckle of his thumb. He raises it to my nose and I stop swaying just long enough to take it in.

Good. Done. Easier than finishing a drink. I don't know why I don't do more drugs. I like cocaine. It's been helpful every once in a great while. Note to self: expand your drug repertoire. Maybe drinking's just gotten to be too much of a habit. Maybe I should try other things more often. This is my new resolution. Quit drinking. Get healthy. Just use drugs. Medicinally, strategically. Be less messy.

I look at the drug boy again and put my finger to my nose, signaling for more. He raises his eyebrows. *Yes, idiot. I want more. Don't fuck with my newfound life here. I've decided from here on out to eat better, exercise regularly, and do more drugs. Give a girl a hand, pal.*

He digs back into his pocket and pulls out the packet. Another bump goes up my nose.

This is perfect. So easy. I should've been doing more drugs all along. I can't believe Jack wouldn't let me try crack the one time I asked. *If you like cocaine,* my inner advertising mind was singing, *you'll LOVE crack!* Who knows? It could've been the solution to all our relationship woes. Both of us on crack. We could've covered for each other. I could've come home from a hard day at the office and he would've had a nice fresh rock cooking in the kitchen for our dinner.

Well. Fuck. Him.

Something doesn't feel right.

Suddenly, I can't find the music. The hooks that moved my body in the right direction seem to be coming at different intervals. I lean farther into the cute boy, trying to get back on track.

There's a buzz. There should be a buzz. Cocaine *is* buzz. Only it's not electric. Not fizzy. More dampening. Flat. I close my eyes, trying to rein things in. Pull it together. I think the music has gone into another room. Someone is pushing me. Is it the cute boy? What cute boy?

Chair.

Floor?

Ground?

Behind my eyes is a separate place. Something's not right. Not cocaine. I need to pull it into one piece. It's all spreading out and I need to keep it here behind my eyes. Think of a word. Find an idea. Concentrate on one thing. One. One. One.

Cactus.

Oh, Jack.

Help me.

Please.

I am on vacation at the beach in Cape Cod and I am six and I am on my stomach and I am on a raft and I am in the ocean. I am my own island. This rolling island. And I am frightened and I want more.

My stepfather is behind me, with the waves rolling past his chest. He will push me again. And again I will ride with the scratchy canvas raft, and the pushing wave, and I will scrape up against the sand on the beach, and I will tell him to do it again. And again.

And here is the wave. And this man, holding my feet, will do it again. Push me with the wave. This man who is new to me. This man who is on his first vacation with his brand-new family. This man who hopes to be less of what he is not to me, and more of what he will one day become. This man who does not share a single gene with me and doesn't recognize anything he's familiar with in my eyes yet is trying to begin to love me nonetheless. And the wave is here. And he lets go of the raft because he's smart enough to know that holding on to someone is not always the same as keeping them close. And this wave, bigger than that wave, pushes me forward faster, more insistent. And the raft buckles under me, folding in half. And I slip forward with the wave now pushing me, just me, pushing me down.

And it is quiet as I fold up underwater. And I am afraid.

I do not like the water—what I can't see underneath. The color by numbers book that my mother bought me sits on the beach on a towel. *What Is It by the Sea?* it asks on its cover, with black outlined creatures and monsters swarming on each page. My mother hopes that by knowing the mysteries of the alien detritus tossed up onto the beach I will be less frightened of the ocean itself. Only I am not. I do not color number 5 brown, and number 7 navy blue. I color in these creatures as I imagine them to be, in their hidden depths. Lime green, bright pink, electric blue.

And now I am with them underneath. Roiling. Flipping. Sinking.

And Jack is sleeping on a beach far from any sea. On sand, cool sand, at the base of a tall saguaro cactus. And all around his still body, the purple desert night floods in. Tired, wanting, needing. He is thirsty for something that is emptying from his cells. Something that he cannot find in the desert. And the stars collect above his head as he sleeps, slowly emptying. And I am not there, either touching him or in his memory. No one is there as he empties.

The saltwater that fills my mouth when I scream does not choke me. It seeps through my body, making me give in, and sink. What Is It by the Sea?

Under the water I am sparkling with the creatures. I feel them swarm around me and I am afraid to open my eyes, but then I do and they are everywhere. Floating, swimming, dancing, swirling. Those creatures of the sea. Darkly sparkling. My sun-blond little boy hair weightlessly waving as I somersault and flail, trying to find the rhythm of the currents that push and pull and bend me. The sea grass and kelp tangle around my legs, tugging, dancing a strange tarantella with me, absent of any music, of any sound other than the bass rush of water entering my lungs.

Jack sleeps on as a river of scorpions and spiders and lizards scamper across his body ahead of the rush of water racing toward him. He does not notice the mineral smell of ozone that precedes the storm coming from the west, or hear the thunder constantly rolling. He is a dry husk, and is lifted up and pushed by the water as it streams in. There is nothing to hold him, to anchor him as he drifts off with the rolling waves, the quiet mariachi music from a faraway village echoing over the rushing water. He is drowning like a fish in the desert.

And hands on my chest. And hands on my legs. And I am back above in the noise of the air and the waves and the beach. And my dad's arms carry me back to shore as the sun bakes the water off my skin, leaving a faint white dusting of salty residue.

I am not crying. I have seen a place for me. Falling sparkles all around. Black and crushing. Crushing and hugging. Liquid and thick. And as deep as one sinks, the sparkles still come from above, from the sun. And I know the way up if up is what I want. When I want it.

Because I am not crying, my new father is laughing.

And my mother, far away on the beach, tan and young and beautiful, is laughing.

And my brother, busy rescuing the runaway raft, is laughing.

And I am Laughing by the Sea.

I am outside the club by the back entrance on Twenty-eighth Street. I drink the club soda that is offered not because I want it but because it is offered nicely. They are all talking and laughing about the times they'd fallen into K-holes.

Ketamine. Special K. Well, at least it wasn't heroin. I'd always thought heroin was such a trashy drug.

Across the street men are loading boxes into a line of UPS trucks, their breath steaming in the morning air. The air is pink and flat, the sun lazy in its arrival. It's snowed during the night—a light dry snow like sand that scatters easily as the busy men walk around their trucks.

Standing up, I thank the strangers for the soda, and for sitting with me, and tell them it's okay, to go back into the club. Which they do.

I'm a little dizzy and I lean against the outside wall of the club. I scoop a handful of fresh white snow off the stair railing next to me and put it in my mouth. It's clear and cold and clean and new.

I try to make myself realize, because it is morning, and everything is white, that I am starting something new and good. But I know this is not true. What is true is only that it is the next day.

I try to make myself realize that I will go back to the penthouse and start packing up my things in a fever of rebirth. But more likely, I will simply watch a little TV and read the paper and order some eggs from the deli. I will maybe make a few calls to see if anyone's looking for a new roommate.

I try to make myself realize that I had beaten up the city. Or that it had beaten *me* up. But instead I know that in the short year we've been together, the city hasn't really noticed me at all. We merely weakly smile and occasionally nod at each other like second cousins at a family reunion.

I try to make myself realize that Jack was and always will be the greatest love of my life. But I know eventually he will be filed under "H" for "hooker" in my expanding file cabinet of funny stories to be pulled out whenever I need a cheap laugh.

But the thing is, right now, he still *is* all that I love, and I don't really have it in me to look for something else to take his place. I am finally, honestly, tired.

I try to make myself realize that I have learned the difference between right and wrong. That there *is such a thing* as right and wrong. But instead I've learned that these are things—this "right," this "wrong"—these are things that we are told. Simply told to believe. These are things we have not tested. And while most of the things we are told may be true, it is not until we have tested them, taunted them, flaunted them, that we truly know they are right. Or wrong. Or true. Or false. Or somewhere in-the-fucking-*between*. And I think I know now a little better which is which. And I also know I'll never quit testing this world. I'll never rely on common knowledge. Or common denominators. Or even common sense, for that matter. To do so would be too, well, *common*.

So. I'll keep dancing in my costumes. Day and night. And I won't sleep as much as I should. And I will drink more than I should. And maybe, as I'm twirling and glittering, playing a retarded game of hide and seek in the middle of an open field, maybe, just maybe, whatever happens next will be bigger, and I will forget that which seems so huge to me right now.

It'll be easier to get a cab on Eleventh Avenue. They line up there waiting for the club to empty. A small group of UPS men watch me gather my bag and straighten myself to my full height. The street is plowed and easier to walk on than the sidewalk.

I don't have a plan yet, but when I do, I know it will be elegant. Yes. *A very elegant plan.*

I wave to the UPS men, and they wave back.

Good morning, boys.

Epilogue

Hugs and Fishes.

Aqua never saw Jack again.

Back in the East Village, she moved in and out of friends' apartments, going out less and less. Fewer people remembered her each time, and that was okay with her. Well not okay, but inevitable, and that passes for okay.

Toward the end, when she grew tired of going out because it was the same "out" over and over again, she would sit and pluck the stray gray hairs cropping up in her wigs.

Mostly, she just drank and watched a lot of TV. But when *Blue's Clues* got a new host, she turned that off as well.

On some late nights, in the deep haze of her vodka, she would dial Jack's pager repeatedly, just to see if anyone would call back. Eventually Jack's ad disappeared from the back of the local gay magazines, and his pager was disconnected. Still, she would dial the number occasionally anyway, along with Ryan's and Grey's and even Trey's. They all had vanished into the city's maze of boxes. Sink or swim. Sink or swim. There is no round-the-clock lifeguard in New York City.

The fish lived, and the fish died. And new fish took their place—though without the constant practice of regular appearances, the newcomers tended to be lackadaisical, even sloppy, with their performances.

She missed Houdini. She missed Mr. Beefeater. She missed stupidity and insanity and danger and fun. She missed the raw truths that bubble up when survival is gambled and mocked.

Four years later, she lived in three boxes marked "Aqua stuff" in a storage facility five blocks from the penthouse she'd once called home for seven months. When she turned thirty, a frozen pipe burst somewhere over the storage unit and dissolved her boxes and wigs and makeup and costumes into an indistinguishable mess. An iridescent glob that smelled faintly of smoke and sweat and cheap perfume. But she wasn't all that upset by the loss, because she knew when she turned thirty that she was never going to become a movie star, and if you are never going to become a movie star, then you may as well just be who you are. With little or no makeup and jewelry.

And so, around four fifteen one Saturday morning, around the time she used to be coming off a shift—she left.

She left because she was left behind—in rotting cardboard boxes and foggy disappearing memories. She left because she wasn't really needed anymore. What was needed was sleep, and a career, and sobriety, and a boyfriend whose vision of the future didn't include someone dead. And all these things showed up eventually, but not for her. She did all the prep work, then got shuffled aside. And a drag queen can survive a lot, but she cannot survive obscurity.

So she abandoned the baubles and shiny things. She pushed aside the wigs and the metallic thongs. She stepped over the piles of her sparkly history and walked completely naked down to the East River, past twenty-four-hour delis and an undercover vice squad officer who once asked her to blow him in his brown sedan. She walked past small huddles of clubgoers stumbling home from places she'd worked at before they were named whatever they were named today.

Expecting her to hail them, cabs slowed as they passed her. She ignored them. Just as she ignored the other drag queens emerging from the dark doorways trying to joke with her, their lipstick smeared and creasing in the corners of their mouths, and their foundation caking and powdering around their eyes.

She ignored the city as well, as it reluctantly shrugged off another night. The lights in the buildings began to slowly blink off as the sun rose, like sequins popping off an evening gown. She ignored the undressing city as it, over time, had come to ignore her. She just stared at her feet as she walked, refusing to look up at a skyline she'd once looked down on from her white penthouse in the sky.

She climbed up and sat on the metal railing overlooking the freezing, roiling East River, and she *wished to God* she had a double vodka before she pushed off and slipped into the inky currents below. Gliding underneath, pulled down under the surface, her limbs danced in the bass rush of the icy river.

And the reflection of the amber lights from the bridges sparkled on the rippling water around her body like thick curling schools of shimmering goldfish as she passed underneath on her way out to sea.

There was no funeral for Aqua. In lieu of flowers, she respectfully requests that you buy a round for the bar.

And if you didn't know Aqua, that's a shame, because everyone who knew her loved her. And everyone who loved her got their fair share back.

And she just wants everyone to remember—please remember—that once there was a darkly sparkling, glittering, shimmering, lovely dangerous time in this city when Aqua loved Jack.

And Jack loved Aqua.

And I loved Jack.

And Jack loved me.

And boys will be boys.

And boys will be girls.

And sometimes the show can't go on.

Acknowledgments

Incalculable thanks to my parents, David and Jackie, for whom I've been, at times, an endurance sport. Without their values I would be nothing more than a very pretty corpse. I apologize for the fearful times; you did not deserve them and will never relive them. Equal gratitude for Rick, my polar opposite yet Siamese twin. For James—WWJD?—and Maya—WDMT? For Arthur and Bob, who came first. For Andy, Milkman Joe Daley, and Clive . . . thank you for looking where one would least expect, and finding something I wasn't sure anyone would want to see. For my editor, Maureen, who's revered by moguls and movie stars, and hardly needed to add an ex–drag queen to her court, but did. And

Stephanie too. And Team Kismet at Harper Perennial. Thank you.

For James and Terrance, Leah, Jeannie, Marty and Jen, Bill and Gilles, the Goddess Raven, Meredith, Jo W., Matty-Patty, Randy, Greg Kadel, John Nathan, Edith, Laura, the "Lady" Bunny, and Jill . . . whatever has become of my life, it's all your fault.

And for every twelve-year-old boy who wore his mother's eyeshadow. To school.

About the author

About the book

Read on

P.S.

Insights,
Interviews
& More...

The Josh Kilmer-Purcell Multiple Choice Quiz

(See below for answers)

1. Josh Kilmer-Purcell was born in:
   (a) Halibut Cove, Alaska
   (b) Ravena, New York
   (c) Kilmer, Kansas

2. When he was eight, his family moved to _____ in a _____:
   (a) Newport, Rhode Island; Learjet
   (b) Beaver Bend, Arkansas; Winnebago "Mini-Winnie"
   (c) Oconomowoc, Wisconsin; Ford Maverick

3. A tight end on his high-school football team, Josh established a single-game state record by catching:
   (a) 17 passes
   (b) 20 passes
   (c) Josh holds no such record

4. Josh Kilmer-Purcell's immediate family are:
   (a) "Taxidermists, goose-calling champions, and retailers of coon pelts, doe-urine lures, and stuff like that"
   (b) "Charter members of the Atheist Alliance"
   (c) "Hardworking, intelligent, well-read, God-fearing, garage-cleaning persons"

Courtesy of the author

5. In tenth grade, Josh moved to:
   (a) Mansfield, Massachusetts
   (b) Provincetown, Massachusetts
   (c) Key West, Florida

6. At college, Josh first majored in:
   (a) Hotel/Restaurant Management
   (b) Bassoon
   (c) History

7. Josh graduated from college and went to work as:
   (a) An intern for Representative Barney Frank
   (b) An assistant store manager at an espresso shop
   (c) An English teacher in Guinea-Bissau, Swaziland, and other developing nations

8. Josh moved to Atlanta in order to:
   (a) Catalog Jane Fonda's private collection of Andy Warhol's portraits of herself
   (b) Attend advertising school
   (c) Sit in for the Principal Bassoon of the Atlanta Symphony Orchestra

9. One of the highlight's of Josh's career came when:
   (a) He won a Clio award for an oatmeal ad featuring a nun, a contortionist, and five caribou
   (b) He ran a successful telethon for Pat Robertson's 700 Club
   (c) He produced two public-service announcements for the Human Rights Campaign featuring Matthew Shepard's mother ▶

The Josh Kilmer-Purcell
Multiple Choice Quiz *(continued)*

10. The idea for Aqua was conceived when Josh:
    (a) Watched the Royal Wedding
    (b) Ducked into the Tennessee Aquarium to escape a gang of rock-throwing homophobes
    (c) Watched film star Esther Williams in the aquacade *Neptune's Daughter*

11. Josh lives with:
    (a) His pet Quentin, an illegal Savannah cat
    (b) His partner, a physician
    (c) Himself and no one else

12. When Josh isn't working at his advertising agency, he's probably:
    (a) Traveling, writing, or visiting Datalounge.com
    (b) Tinkering with his 1966 Jaguar XKE or modeling his hands for Palmolive
    (c) Breeding bigger, tougher goldfish for his Aqua comeback

Answers:

1. *(b): "I was born in upstate New York, near Albany, where my family had lived for generations, having scuttled over from Massachusetts in the mid-1600s. I mention New England to better disguise my extraction from a long line of Catskill hillbillies."*

2. *(c) "Joining me in the 1976 Maverick were my older brother, mother, new stepfather, golden lab, cat, pots, pans, and my vast*

collection of Richie Rich *comics. The
name of our town, Oconomowoc, is Native
American for 'where the waters meet' or
'pale boy is squaw wannabe.' Oconomowoc
is a dichotomous community, divided into
wealthy 'lake people' and farmers, of which
we were neither. I grew up in a small ranch
house, in a five-home subdivision, miles
out of town, completely surrounded by
cornfields. Oconomowoc had once been a
summer community for Chicago business
magnates—Oscar Meyer, the Pabsts, the
Swansons (of Hungry Man Dinners fame).
Rich kids and rural kids went to the same
schools, and mingled together with
surprisingly little difficulty."*

3. (c): *"The fact that I played bassoon in our
regionally renowned school band and
wasn't mocked mercilessly is proof of my
town's great midwestern socialist ethic."*

4. (c): *"My very existence disproves any
credible theory of genetics."*

5. (a): *"I had dreams of finding my aristocratic
roots when I moved back East. Instead, I worked
in a feed-and-grain warehouse, which burned to
the ground one day. They evacuated a five-
square-mile radius around the blaze because of
the amount of toxic pesticides stored inside.
Mansfield was a blue-collar suburb of Boston,
when its citizens bothered to wear collars. I hid
from the masses in a small clique of college-bound
nerds, then escaped back to the Midwest* ▸

to attend Michigan State University.
Nowadays I hear Mansfield is quite posh.
I like to think I started the trend."

6. *(a):* "*I soon discovered the Hotel/Restaurant*
program required skills beyond tasting
new menu items and hiring bell boys.
Accordingly, I transferred into the English
department. My collegiate experience had
its moments. I wrote and illustrated a
mildly popular cartoon for the student
paper. Cartooning helped me realize that
the only innate talent I had was making
fun of others. Also at MSU, I took classes
taught by Diane Wakoski, a legendary poet
whose work made me feel wholly talentless
by comparison. I urge aspiring writers to
avoid her poems at all costs: read them, and
you'll come to inevitably hate your own
pathetic scribblings; next you'll wake up on
a subway, dressed in drag, drunk as a lord."

7. *(b):* "*I worked in the local East Lansing*
espresso bar, then a rarity. Eventually, the
owner wanted me to go into business with
her, spreading lattes across the Midwest.
'Who,' I thought, 'would pay four dollars for
a coffee?' "

8. *(b):* "*After my ill-suited stint in the service*
industry, I moved to Atlanta to study at
Portfolio Center, a sort of advertising
vocational school. I'd always had a bent for
lying; advertising would give me the stage
I deserved. So I enrolled in the art direction
curriculum, convinced that copywriters
had to do all the work. How difficult could it

be to pick out a typeface? I toiled harder at
Portfolio Center than I had in all my life.
It's a skill few master: lying persuasively to
different demographic markets. My career
decision paid off, though. Many awards
later, my karmic consolation is that there
are tens of thousands of people out there
drinking Absolut and wearing what they
believe to be whiter whites."

9. (c): "Judy Shepard is the closest I've come
to an actual saint. I still cry when I read
her and her husband's court statement that
spared the death penalty for Matthew's
killers. Their bravery is beyond measure."

10. (a): "I was an avid TV junkie, and dutifully
roused everyone in my summer camp at four
a.m. the day of Princess Diana's wedding.
While watching the bucketfuls of pomp
on a creaky wooden bench in the cafeteria,
I realized that if a simple, rather homely
kindergarten teacher could become a global
spectacle, then there was hope for me to
become . . . princess! . . . of! . . . the! . . .
world!"

11. (b): "Yes, I now live with my partner of six
years on New York's Upper East Side. He
saved my life, as I'm clearly not fit to
undertake such a massive task on my own.
He's a doctor, which makes my parents very
happy, and he recently earned his MBA as
well. He's 'going places,' as they say, which
hopefully doesn't lead him someplace
I'm not." ▶

The Josh Kilmer-Purcell
Multiple Choice Quiz *(continued)*

12. *(a):* "*My partner and I spend the time
between our petty arguments traveling
the world. He favors roughing it, I favor
rouging it. So we alternate between grueling
explorations in the African bush and jaunts
to southern France to visit my elderly gay
uncle and his partner, who've been together
since the 1950s. If they were ever gauche
enough to sell their stories, I wouldn't bother
telling mine. At home, I can usually be found
stealing ideas from the incomparably witty
posters at Datalounge.com. They've become
my virtual barmates, since, in gay years,
I'm nearing eighty-five and my ego cannot
withstand the trendier clubs anymore.
Nor can my liver, according to my doctor
boyfriend (I did mention that he's a doctor,
no?). I don't watch TV at home or anywhere
else. Neither should you. Except for the ads,
of course. Oh, I almost forgot: I'm working
on a second book, which will likely cement
my status as a one-hit wonder.*" ⟋

Aqua Bound
The Queer Path to Publication

WHILE THE STORY OF AQUA, myself, and Jack might seem surreal, the story of how it became a book is not without peculiarities of its own.

I'd been writing short Aqua stories for my own amusement ever since I'd packed her away. It was a way to hang on to her in some form, even though, for my own survival, I'd had to move on. True to character, she refused to fade into oblivion.

Tucked away in the advertising world, I've had the fortune to work as art director alongside many great copywriters. One of my favorites was Maya Frey, wife of James Frey, author of *A Million Little Pieces* and *My Friend Leonard.* She's an incredibly talented fiction writer in her own right, ▶

Aqua photograph credits:
Photograph by Greg Kadel
Retouching by Matthew Baldwin
Hair by Andre Gunn
Makeup by Mariel Barrera
Agents and Managers by Jason Sharpe, Jo Weinberg, and Marek Milewicz

> **❝ True to character, [Aqua] refused to fade into oblivion. ❞**

Aqua Bound *(continued)*

Once, during a shoot in Los Angeles, Maya and I decided to ritualize our haphazardly executed creative writings habits. We set ourselves deadlines, and typed away every spare moment. One time during my newfound zealotry, I spent a wildly productive two weeks at James and Maya's beach house on Amagansett, Long Island. It was mid-January. I had no car, no company, and no more to eat than I'd managed to carry on the train.

Eight months after languishing poolside in Los Angeles and bewailing my indiscipline as an artist, I had a rough manuscript. "It doesn't suck," pronounced James. More encouraging yet, he helped me submit the manuscript to several agents. One agent, Andy McNicol, took the bait.

Then, serendipity took over.

Satisfied that I had suckered an agent into taking me on, I flew to France on vacation, expecting nothing more than a stack of rejection letters upon my return.

My last night in Nice, I pegged a quiet pool-goer for an American—in that way none but a fellow American can do—and, though we hadn't spoken, he approached me the following day at the airport. "How did you like the hotel?" he asked.

His name was Joe Daley, and he was the first star my own had crossed. We sat next to each other on the plane, and I discovered he was an executive producer for Clive Barker's film production company. I foisted my manuscript on him, hoping I'd found another sucker.

(A short aside: my best friend, ex-roommate, and early Aqua enthusiast

> 66 My last night in Nice, I pegged a quiet pool-goer for an American—in that way none but a fellow American can do. 99

from back in Atlanta was a huge Clive Barker fan. He had every first edition, signed movie poster, etc. John Wright died in 1999, but I've no doubt each twist of fate encountered by this book can be attributed to him.)

Back in Los Angeles, Joe read and passed the book on to Clive, who in turn passed it on to Maureen O'Brien, a senior executive editor at HarperCollins. "Don't hold your breath," I was told. "If she likes it, maybe, just maybe, she'll pass it on to a junior editor."

I figured I had reached my limit of attracting suckers.

But for whatever reason, my streak continued. Maureen read it and took it on as her own. Her French bulldog, Mimi, snarled at anyone who approached it threateningly.

From a drunken night in Los Angeles, to a man on a plane, to a French bulldog named Mimi, I've just stood back and watched everything unfold, immensely grateful for it all. ᕦ

> 66 Maureen O'Brien's French bulldog, Mimi, snarled at anyone who approached my manuscript threateningly. 99

Finding Jack

I DIDN'T HAVE ANY IDEA what had happened to Jack after the day he left for the desert.

I never reached out in those early days after we parted. I was cowardly, probably for my own good. But not for my own peace of mind. For the first year or so after we separated, I was comforted simply to see his escort ad in the back pages of the local gay rags.

When that disappeared, and his pager number was disconnected, I had no way to reach him.

In the midst of editing this manuscript, my curiosity overcame my fear. It had taken me nine years of not knowing, and twenty-seven chapters of wondering, to get to that point.

According to the Internet, there are eighteen people with Jack's last name in the area of Southern California where he'd grown up. I figured one of them must either be a relative, or possibly one of his parents. I no longer remembered their first names.

Sitting in an easy chair on a Friday afternoon, in the apartment I share with my partner of six years, I started down the phone list. I left stilted messages on answering machines, and apologized to people who had never heard of a Jack F_____.

I had a speech prepared for "Oh sure, here's his phone number." I had another speech prepared for "Jack is dead."

I was not prepared for what lay between those contingencies.

GORDON F, *his voice old, gravelly, cautious:*
"My son is Jack. He has an unlisted number."

> 66 For the first year or so after Jack and I separated, I was comforted simply to see his escort ad in the back pages of the local gay rags. 99

ME: "I was a good friend of Jack's in New York. . . . Could you pass on a message for me?"

GORDON F: "I just got off the phone with him. I can call him on my cell. Hang on."

[*Sound of the cell phone being dialed.*]

GORDON F (to JACK): "Hey, a person named Josh is on the phone. He says he knew you in New York."

GORDON F (to ME): "He wants to know how Aqua is."

ME, *laughing:* "Oh, Aqua's long gone."

GORDON F (to JACK): "He says Aqua's not around anymore."

GORDON F (to ME): "Jack wants to know what happened to her."

ME, *laughing nervously:* "She just retired. That's all."

GORDON F (to JACK): "She retired. . . . Okay, hang on."

GORDON F (to ME): "He wants to know what you want."

ME: "Just tell Jack that I've been thinking about him and wish him the best. And that I've written a book, and I thought he might be interested in talking about it."

GORDON F (to JACK): "He's been thinking about you. And he's writing a book. Wants to know if you want to talk."

GORDON F (to ME): "Jack wishes you luck."

ME: "Okay. I just wanted to catch up . . . would you maybe give him my phone number in case he wants to get a hold of me?" ▶

6 6 'I was a good friend of Jack's in New York. . . . Could you pass on a message for me?' 9 9

6 6 'Just tell Jack that I've been thinking about him.' 9 9

Finding Jack *(continued)*

GORDON F (to JACK): "He wants to leave his number."

GORDON F (to ME): "Jack says he will find you."

Jack hasn't found me yet. And I guess I have to be okay with that. I'm no longer all that lost anyway. Maybe that's what I would like him to see. That his *was* an elegant plan after all.
I am very happy that he is alive.

Azucar, Jack ⌒

Music to Read
Memoirs By

1. The theme song to "The Bob Newhart Show"—a jazz classic for anyone who gets up each day and goes "into the office." It will put a bounce in your step as you march along . . . in the rain . . . in your miserable rut.

2. "Mad World," Gary Jules—for the disaffected teen in all of us.

3. "Slung-lo," Erin McKeown—*"I'm turning this B-side around to a de-light. . . . I'm dancing till I drop. Oh, one small step, first right, now left, I'm never gonna stop."*

4. "Ma Vlast/My Country," Bedrich Smetana—it's a biopic soundtrack waiting to happen.

5. "Light and Day/Reach for the Sun," Polyphonic Spree—inescapable happy goose bumps.

6. "Born Slippy," Underworld— personally, I think this is a slam-dunk tune for a funeral. I'm probably alone in this.

7. Anything by either Rufus or Martha Wainwright—but you already know that.

8. "The Show Must Go On," Dinah Washington—if you insist, Dinah.

9. "Poetry Man," Phoebe Snow—*you make things all rhyme.*

10. "Here's to Life," Shirley Horn—any reason to toast.

11. "Tabula Rasa," Arvo Part—I'm not saying you *will* breakup with your spouse if you listen to this at four a.m., but it will make you consider it.

12. "I'll Plant My Own Tree," Helen Lawson—that's *"The Incomparable Helen La Lawson"* to you, halfpint.

13. "Both Sides Now," Joni Mitchell (off *Dreamland*)—it's not the version you're thinking of. You must be as tall as this sign to listen to it.

14. "Wake up in New York," Craig Armstrong—I have only listened to this song all the way through twice, because it hits torturously square on the mark.

15. "Look at Me," Geri Halliwell—yes, *that* Geri Halliwell. I'm a goddamn homo, dammit. Sue me.

16. "Fancy," Bobbie Gentry—perfect for nine-year-olds trying on their mother's shoes while wearing a floor-length gown made of a bath towel held together by clothespins.

17. "Mr. Lucky Man," by the Kennedys—yes. Yes I am.

18. "Closer to Mercury," Wheat— *"And I would have walked behind. And I would have walked beside you. And I would have told you every lie. And I would've done that for you. Open your eyes sometime. It's funny how I adore you. I'm gonna get every line just fine. I'll even sing it for you now. . . . And there was a time I felt the morning sun rose up for you. And I would've whispered in your ear. And I would make coffee for you. And I would've crossed that line. And I could've drawn it for you. Tell me that everything is fine. Couldn't you do that for me now?"*

Don't miss the next book by your favorite author. Sign up now for AuthorTracker by visiting www.AuthorTracker.com.

– THE FABULOUS –
BEEKMAN BOYS

They're successful city slickers who know nothing about
goat farming. So naturally they buy a goat farm.
Tune in and see what happens next.

Check your local listings

**planet
green** ™

planetgreen.channelfinder.net